P9-DYP-181

10/00

refer

6/17

Lexile: _____

AR/BL: _____

AR Points: _____

experiment
CENTRAL

A-Ec volume 1

experiment
CENTRAL

understanding scientific principles through projects

John T. Tanacredi & John Loret, General Editors

AN IMPRINT OF THE GALE GROUP

DETROIT · NEW YORK · SAN FRANCISCO
LONDON · BOSTON · WOODBRIDGE, CT

experiment
CENTRAL Understanding Scientific Principles Through Projects

Researched, developed, and illustrated by **Book Builders Incorporated**

John T. Tanacredi, *General Editor*
John Loret, *General Editor*

U•X•L Staff

Allison McNeill, *U•X•L Senior Editor*
Elizabeth Shaw, *U•X•L Associate Editor*
Carol DeKane Nagel, *U•X•L Managing Editor*
Thomas L. Romig, *U•X•L Publisher*
Meggin Condino, *Senior Analyst, New Product Development*

Shalice Shah-Caldwell, *Permissions Associate (Pictures)*

Rita Wimberley, *Senior Buyer*
Evi Seoud, *Assistant Production Manager*
Dorothy Maki, *Manufacturing Manager*
Mary Beth Trimper, *Production Director*

Eric Johnson, Tracey Rowens, *Senior Art Directors*

Pamela A. Reed, *Imaging Coordinator*
Christine O'Bryan, *Graphic Specialist*
Randy Basset, *Image Database Supervisor*
Barbara Yarrow, *Graphic Services Manager*

Linda Mahoney, LM Design, *Typesetting*

Library of Congress Cataloging-in-Publication Data.

Loret, John.
    Experiment central: understanding scientific principles through projects / John Loret,
John T. Tanacredi.
        p. cm.
    Includes bibliographical references and index.
    Contents: v. 1. A-Ec — v. 2. El-L — v. 3. M-Sc — v. 4. So-Z
    Summary: Demonstrates scientific concepts by means of experiments, including
    step-by-step instructions, lists of materials, troubleshooter's guide, and interpretation and
    explanation of results.
    ISBN 0-7876-2892-1 (set). — ISBN 0-7876-2893-X (v. 1) — ISBN 0-7876-2894-8 (v.2)
    — ISBN 0-7876-2895-6 (v.3) — ISBN 0-7876-2896-4 (v. 4)
    1. Science-Experiments-Juvenile literature. [1. Science-Experiments. 2.
    Experiments.] I. Tanacredi, John T. II. Title.
Q164 .L57 2000
507'.8-dc21   99-054142

contents

Reader's Guide xiii

Parent's and Teacher's Guide xvii

Experiments by Scientific Field xix

Words to Know xxvii

Volume 1: A-Ec

1. Acid Rain 1: 1
   Acid Rain and Animals: How does acid rain
   affect brine shrimp? 1: 4
   Acid Rain and Plants: How does acid rain
   affect plant growth? 1: 10
2. Annual Growth 1: 19
   Tree Growth: What can be learned from the
   growth patterns of trees? 1: 24
   Lichen Growth: What can be learned from
   the environment by observing lichens? 1: 28
3. Biomes 1: 35
   Building a Temperate Forest Biome 1: 39
   Building a Desert Biome 1: 42
4. Cells 1: 49
   Investigating Cells: What are the differences
   between a multicellular organism and a
   unicellular organism? 1: 52
   Plant Cells: What are the cell differences
   between monocot and dicot plants? 1: 55

contents

5. Chemical Energy 1: 61
   Rusting: Is the chemical reaction exothermic,
   endothermic, or neither? 1: 64
   Exothermic or Endothermic: Determining
   whether various chemical reactions are exothermic
   or endothermic 1: 68
6. Chemical Properties 1: 75
   Slime: What happens when white glue and
   borax mix? 1: 79
   Chemical Reactions: What happens when
   mineral oil, water, and iodine mix? 1: 84
7. Chlorophyll 1: 91
   Plant Pigments: Can pigments be separated? 1: 93
   Response to Light: Do plants grow differently
   in different colors of light? 1: 98
8. Composting/Landfills 1: 105
   Living Landfill: What effect do the
   microorganisms in soil have on the
   decomposition process? 1: 109
   Composting: Using organic material to grow plants 1: 115
9. Density and Buoyancy 1: 123
   Density: Can a scale of relative density predict
   whether one material floats on another? 1: 126
   Buoyancy: Does water pressure affect buoyancy? 1: 132
10. Dissolved Oxygen 1: 139
    Decay and Dissolved Oxygen: How does the
    amount of decaying matter affect the level of
    dissolved oxygen in water? 1: 144
    Goldfish Breath: How does a decrease in the
    dissolved oxygen level affect the breathing
    rate of goldfish? 1: 149
11. Earthquakes 1: 159
    Detecting an Earthquake: How can movement
    of Earth's crust be measured? 1: 162
    Earthquake Simulation: Is the destruction greater
    at the epicenter? 1: 167
12. Eclipses 1: 175
    Simulating Solar and Lunar Eclipses 1: 178
    Phases of the Moon: What does each phase
    look like? 1: 181

experiment
CENTRAL

Budget Index lvii
Level of Difficulty Index lxiii
Timetable Index lxix
General Subject Index lxxv

contents

Volume 2: EI-L

13. Electricity 2: 185
    Electrolytes: Do some solutions conduct electricity
    better than others? 2: 188
    Batteries: Can a series of homemade
    electric cells form a "pile" strong enough to
    match the voltage of a D-cell battery? 2: 193
14. Electromagnetism 2: 203
    Magnetism: How can a magnetic field be
    created and detected? 2: 206
    Electromagnetism: How can an electromagnet
    be created? 2: 210
15. Enzymes 2: 217
    Finding the Enzyme: Which enzyme breaks
    down hydrogen peroxide? 2: 220
    Tough and Tender: Does papain speed up
    the aging process? 2: 224
16. Erosion 2: 231
    Erosion: Does soil type affect the amount of
    water that runs off a hillside? 2: 233
    Plants and Erosion: How do plants affect the
    rate of soil erosion? 2: 239
17. Flight 2: 249
    Lift-Off: How can a glider be made to fly higher? 2: 252
    Helicopters, Propellers, and Centripetal Force:
    Will it fly high? 2: 256
18. Germination 2: 263
    Effects of Temperature on Germination:
    What temperatures encourage and
    discourage germination? 2: 266
    Comparing Germination Times: How fast
    can seeds grow? 2: 270
19. Gravity 2: 277
    Gravity: How fast do different objects fall? 2: 280
    Measuring Mass: How can a balance be made? 2: 285

contents

20. Greenhouse Effect **2**: 291
    Creating a Greenhouse: How much will
    the temperature rise inside a greenhouse? **2**: 294
    Fossil Fuels: What happens when
    fossil fuels burn? **2**: 300
21. Groundwater Aquifers **2**: 307
    Aquifers: How do they become polluted? **2**: 311
    Groundwater: How can it be cleaned? **2**: 316
22. Heat **2**: 323
    Conduction: Which solid materials are
    the best conductors of heat? **2**: 327
    Convection: How does heat move through liquids? **2**: 334
23. Life Cycles **2**: 341
    Tadpoles: Does temperature affect the rate at
    which tadpoles change into frogs? **2**: 343
    Insects: How does food supply affect the growth
    rate of grasshoppers or crickets? **2**: 349
24. Properties of Light **2**: 357
    Looking for the Glow: Which objects glow
    under black light? **2**: 360
    Refraction and Defraction: Making a rainbow **2**: 363

Budget Index lvii
Level of Difficulty Index lxiii
Timetable Index lxix
General Subject Index lxxv

Volume 3: M-Sc

25. Magnetism **3**: 369
    Magnets: How do heat, cold, jarring, and
    rubbing affect the magnetism of a nail? **3**: 373
    Electromagnets: Does the strength of an
    electromagnet increase with greater current? **3**: 379
26. Microorganisms **3**: 387
    Microorganism: What is the best way
    to grow penicillin? **3**: 389
    Growing Microorganisms in a Petri Dish **3**: 394
27. Mixtures and Solutions **3**: 403
    Suspensions and Solutions: Can filtration
    and evaporation determine whether mixtures
    are suspension or solutions? **3**: 406

experiment
CENTRAL

Colloids: Can colloids be distinguished from
suspension using the Tyndall effect? 3: 411
28. Nutrition 3: 419
Energizing Foods: Which foods contain
carbohydrates and fats? 3: 423
Nutrition: Which foods contain proteins and salts? 3: 426
29. Optics/Optical Illusions 3: 431
Optics: What is the focal length of a lens? 3: 432
Optical Illusions: Can the eye be fooled? 3: 437
30. Osmosis/Diffusion 3: 445
Measuring Membranes: Is a plastic bag a
semipermeable membrane? 3: 448
Changing Concentrations: Will a bag of salt
water draw in fresh water? 3: 453
31. Oxidation-Reduction 3: 461
Reduction: How will acid affect dirty pennies? 3: 463
Oxidation and Rust: How is rust produced? 3: 469
32. pH 3: 477
Kitchen Chemistry: What is the pH of
household chemicals? 3: 481
Kitchen Chemistry: Chemical Titration:
What is required to change a substance from
an acid or a base into a neutral solution? 3: 486
33. Photosynthesis 3: 493
Photosynthesis: How does light affect plant growth? 3: 496
Light Intensity: How does the intensity of
light affect plant growth? 3: 501
34. Potential and Kinetic Energy 3: 509
Measuring Energy: How does the height of an
object affect its potential energy? 3: 512
Using Energy: Build a roller coaster 3: 517
35. Rocks and Minerals 3: 527
Mineral Testing: What kind of mineral is it? 3: 531
Rock Classification: Is it igneous, sedimentary
or metamorphic? 3: 534
36. Salinity 3: 541
Making a Hydrometer: How can salinity
be measured? 3: 544
Density Ball: How to make a standard for
measuring density 3: 550

37. Scientific Method 3: 557
    Using the Scientific Method: What are the
    mystery powders? 3: 562
    Using the Scientific Method: Do fruit flies
    appear out of thin air? 3: 568

Budget Index lvii
Level of Difficulty Index lxiii
Timetable Index lxix
General Subject Index lxxv

Volume 4: So-Z

38. Solar Energy 4: 575
    Capturing Solar Energy: Will seedlings grow
    bigger in a greenhouse? 3: 579
    Solar Cells: Will sunlight make a motor run? 3: 584
39. Sound 4: 589
    Wave Length: How does the length of a
    vibrating string affect the sound it produces? 3: 592
    Pitch: How does the thickness of a vibrating
    string affect sound? 3: 596
40. Stars 4: 603
    Tracking Stars: Where is Polaris? 3: 606
    Tracking the Motion of the Planets: Can a
    planet be followed? 3: 609
41. Static Electricity 4: 615
    Building an Electroscope: Which objects are
    electrically charged? 3: 619
    Measuring a Charge: Does nylon or wool
    create a stronger static electric charge? 3: 625
42. Structures and Shapes 4: 633
    Arches and Beams: Which is strongest? 3: 637
    Beams and Rigidity: How does the vertical
    height of a beam affect its rigidity? 3: 640
43. Tropisms 4: 647
    Phototropism: Will plants follow a
    maze to reach light? 3: 650
    Geotropism: Will plant roots turn toward
    the pull of gravity? 3: 655

44. Vegetative Propagation 4: 665
    Auxins: How do auxins affect plant growth? 3: 668
    Potatoes from Pieces: How do potatoes
    reproduce vegetatively? 3: 676
45. Volcanoes 4: 683
    Model of a Volcano: Will it blow its top? 3: 687
    Looking at a Seismograph: Can a volcanic
    eruption be detected? 3: 691
46. Properties of Water 4: 697
    Cohesion: Can the cohesive force of surface tension
    in water support an object denser than water? 3: 701
    Adhesion: How much weight is required to break
    the adhesive force between an object and water? 3: 705
47. Rivers of Water 4: 713
    Weathering Erosion in Glaciers: How does a river
    make a trench? 3: 717
    Stream Flow: Does the stream meander? 3: 721
48. Water Cycle 4: 729
    Temperature: How does temperature affect the
    rate of evaporation? 3: 731
    Surface Area: How does surface area affect the
    rate of evaporation? 3: 736
49. Weather 4: 745
    Wind: Measuring wind speed with a
    homemade anemometer 3: 748
    Clouds: Will a drop in air temperature
    cause a cloud to form? 3: 752
50. Weather Forecasting 4: 759
    Dewpoint: When will dew form? 3: 764
    Air Pressure: How can air pressure be measured? 3: 769

Budget Index lvii
Level of Difficulty Index lxiii
Timetable Index lxix
General Subject Index lxxv

reader's guide

Experiment Central: Understanding Scientific Principles Through Projects provides in one resource a wide variety of experiments covering nine key science curriculum fields—Astronomy, Biology, Botany, Chemistry, Ecology, Geology, Meteorology, Physics, and Scientific Method—spanning the earth sciences, life sciences, and physical sciences.

One hundred experiments and projects for students are presented in 50 subject-specific chapters. Chapters, each devoted to a scientific concept, include: Acid Rain, Biomes, Chemical Energy, Flight, Greenhouse Effect, Optics, Solar Energy, Stars, Volcanoes, and Weather. Two experiments or projects are provided in each chapter.

Entry format

Chapters are arranged alphabetically by scientific concept and are presented in a standard, easy-to-follow format. All chapters open with an explanatory overview section designed to introduce students to the scientific concept and provide the background behind a concept's discovery or important figures who helped advance the study of the field.

Each experiment is divided into eight standard sections designed to help students follow the experimental process clearly from beginning to end. Sections are:

- Purpose/Hypothesis
- Level of Difficulty
- Materials Needed
- Approximate Budget

- Timetable
- Step-by-Step Instructions
- Summary of Results
- Change the Variables

Each chapter also includes a "Design Your Own Experiment" section that allows students to apply what they have learned about a particular concept and create their own experiments. This section is divided into:

- How to Select a Topic Relating to this Concept
- Steps in the Scientific Method
- Recording Data and Summarizing the Results
- Related Projects

Concluding all chapters is a "For More Information" section that provides students with a list of books with further information about that particular topic.

Special Features

- A "**Words to Know**" section runs in the margin of each chapter providing definitions of terms used in that chapter. Terms in this list are bolded in the text upon first usage. A cumulative glossary collected from all "Words to Know" sections in the 50 chapters is included in the beginning of each volume.

- **Experiments by Scientific Field** index categorizes all 100 experiments by scientific curriculum area.

- **Parent's and Teacher's Guide** recommends that a responsible adult always oversee a student's experiment and provides several safety guidelines for all students to follow.

- Standard sidebar boxes accompany experiments and projects:

  "**What Are the Variables?**" explains the factors that may have an impact on the outcome of a particular experiment.

  "**How to Experiment Safely**" clearly explains any risks involved with the experiment and how to avoid them. While all experiments have been constructed with safety in mind, it is always recommended to proceed with caution and work under adult supervision while performing any experiment (please refer to Parent's and Teacher's Guide on page xvii).

  "**Troubleshooter's Guide**" presents problems that a student might encounter with an experiment, possible causes of the problem, and ways to remedy the problem.

- **Budget Index** categorizes experiments by approximate cost. Budgets may vary depending on what materials are readily available in the average household.

- **Level of Difficulty Index** lists experiments according to "Easy," "Moderate," "Difficult," or combination thereof. Level of difficulty is determined by such factors as the time necessary to complete the experiment, level of adult supervision recommended, and skill level of the average student. Level of difficulty will vary depending on the student. A teacher or parent should always be consulted before any experiment is attempted.

- **Timetable Index** categorizes each experiment by the time needed to complete it, including set-up and follow-through time. Times given are approximate.

- **General Subject Index** provides access to all major terms, people, places, and topics covered in *Experiment Central*.

- Approximately **150 photographs** enhance the text.

- Approximately **300 drawings** illustrate specific steps in the experiments, helping students follow the experimental procedure.

Acknowledgments

Credit is due to the general editors of *Experiment Central* who lent their time and expertise to the project, and oversaw compilation of the volumes and their contents:

John T. Tanacredi, Ph.D.
Adjunct Full Professor of Ecology
Department of Civil and Environmental Engineering,
Polytechnic University
Adjunct Full Professor of Environmental Sciences,
Nassau Community College, State University of New York
President, The Science Museum of Long Island

John Loret, Ph.D., D.Sc.
Professor Emeritus and Former Director of Environmental
Studies of Queens College, City University of New York
Director, The Science Museum of Long Island

A note of appreciation is extended to the *Experiment Central* advisors, who provided their input when this work was in its formative stages:

Comments and Suggestions

We welcome your comments on *Experiment Central.* Please write: Editors, *Experiment Central,* U•X•L, 27500 Drake Rd., Farmington Hills, Michigan, 48331–3535; call toll free: 1–800–877–4253; fax: 248–414–5043; or send e-mail via http://www.galegroup.com.

experiment
CENTRAL

parent's and teacher's guide

The experiments and projects in *Experiment Central* have been carefully constructed with issues of safety in mind, but your guidance and supervision are still required. Following the safety guidelines that accompany each experiment and project (found in the "How to Experiment Safely" sidebar box), as well as putting to work the safe practices listed below, will help your child or student avoid accidents. Oversee your child or student during experiments, and make sure he or she follows these safety guidelines:

- Always wear safety goggles if there is any possibility of sharp objects, small particles, splashes of liquid, or gas fumes getting in someone's eyes.

- Always wear protective gloves when handling materials that could irritate the skin.

- Never leave an open flame, such as a lit candle, unattended. Never wear loose clothing around an open flame.

- Follow instructions carefully when using electrical equipment, including batteries, to avoid getting shocked.

- Be cautious when handling sharp objects or glass equipment that might break. Point scissors away from you and use them carefully.

- Always ask for help in cleaning up spills, broken glass, or other hazardous materials.

- Always use protective gloves when handling hot objects. Set them down only on a protected surface that will not be damaged by heat.

parent's and teacher's guide

- Always wash your hands thoroughly after handling material that might contain harmful microorganisms, such as soil and pond water.

- Do not substitute materials in an experiment without asking a knowledgeable adult about possible reactions.

- Do not use or mix unidentified liquids or powders. The result might be an explosion or poisonous fumes.

- Never taste or eat any substances being used in an experiment.

- Always wear old clothing or a protective apron to avoid staining your clothes.

experiments by scientific field

Astronomy

[Eclipses] Simulating Solar and Lunar Eclipses **1**: 178

[Eclipses] Phases of the Moon: What does each
phase look like? **1**: 181

[Stars] Tracking Stars: Where is Polaris? **3**: 606

[Stars] Tracking the Motion of the Planets: Can a
planet be followed? **3**: 609

Chapter name in brackets, followed by experiment name; **bold type** indicates volume number, followed by page number.

Biology

[Cells] Investigating Cells: What are the differences
between a multicellular organism
and a unicellular organism? **1**: 52

[Cells] Plant Cells: What are the cell differences
between monocot and dicot plants? **1**: 55

[Composting/Landfills] Living Landfill: What effect
do the microorganisms in soil have on the
decomposition process? **1**: 109

[Composting/Landfills] Composting: Using organic
material to grow plants **1**: 115

[Enzymes] Finding the Enzyme: Which enzyme breaks
down hydrogen peroxide? **2**: 220

[Enzymes] Tough and Tender: Does papain speed up the
aging process? **2**: 224

[Life Cycles] Tadpoles: Does temperature affect the
rate at which tadpoles change into frogs? **2**: 343

experiments by scientific field

[Life Cycles] Insects: How does food supply affect the growth rate of grasshoppers or crickets? 2: 349

[Microorganisms] Microorganism: What is the best way to grow penicillin? 3: 389

[Microorganisms] Growing Microorganisms in a Petri Dish 3: 394

[Nutrition] Energizing Foods: Which foods contain carbohydrates and fats? 3: 423

[Nutrition] Nutrition: Which foods contain proteins and salts? 3: 426

[Osmosis and Diffusion] Measuring Membranes: Is a plastic bag a semipermeable membrane? 3: 448

[Osmosis and Diffusion] Changing Concentrations: Will a bag of salt water draw in fresh water? 3: 453

Botany

[Annual Growth] Tree Growth: What can be learned from the growth patterns of trees? 1: 24

[Annual Growth] Lichen Growth: What can be learned from the environment by observing lichens? 1: 28

[Chlorophyll] Plant Pigments: Can pigments be separated? 1: 93

[Chlorophyll] Response to Light: Do plants grow differently in different colors of light? 1: 98

[Germination] Effects of Temperature on Germination: What temperatures encourage and discourage germination? 2: 266

[Germination] Comparing Germination Times: How fast can seeds grow? 2: 270

[Photosynthesis] Photosynthesis: How does light affect plant growth? 3: 496

[Photosynthesis] Light Intensity: How does the intensity of light affect plant growth? 3: 501

[Tropisms] Phototropism: Will plants follow a maze to reach light? 3: 650

[Tropisms] Geotropism: Will plant roots turn toward the pull of gravity? 3: 655

[Vegetative Propagation] Auxins: How do auxins affect plant growth? 3: 668

[Vegetative Propagation] Potatoes from Pieces: How do potatoes reproduce vegetatively? 3: 676

Chemistry

[Chemical Energy] Rusting: Is the chemical reaction
exothermic, endothermic, or neither? 1: 64

[Chemical Energy] Exothermic or Endothermic:
Determining whether various chemical reactions are
exothermic or endothermic 1: 68

[Chemical Properties] Slime: What happens when white
glue and borax mix? 1: 79

[Chemical Properties] Chemical Reactions: What happens
when mineral oil, water, and iodine mix? 1: 84

[Mixtures and Solutions] Suspensions and Solutions:
Can filtration and evaporation determine whether
mixtures are suspension or solutions? 3: 406

[Mixtures and Solutions] Colloids: Can colloids be
distinguished from suspension using the
Tyndall effect? 3: 411

[Oxidation-Reduction] Reduction: How will acid affect
dirty pennies? 3: 463

[Oxidation-Reduction] Oxidation and Rust: How is
rust produced? 3: 469

[pH] Kitchen Chemistry: What is the pH of
household chemicals? 3: 481

[pH] Chemical Titration: What is required to
change a substance from an acid or a base
into a neutral solution? 3: 486

[Salinity] Making a Hydrometer: How can salinity
be measured? 3: 544

[Salinity] Density Ball: How to make a standard for
measuring density 3: 550

[Properties of Water] Cohesion: Can the cohesive force
of surface tension in water support an object denser
than water? 3: 701

[Properties of Water] Adhesion: How much weight is
required to break the adhesive force between an object
and water? 3: 705

Ecology

[Acid Rain] Acid Rain and Animals: How does acid rain
affect brine shrimp? 1: 4

[Acid Rain] Acid Rain and Plants: How does acid rain
affect plant growth? 1: 10

**experiments
by scientific
field**

experiments by scientific field

[Biomes] Building a Temperate Forest Biome 1: 39

[Biomes] Building a Desert Biome 1: 42

[Composting/Landfills] Living Landfill: What effect
do the microorganisms in soil have on the
decomposition process? 1: 109

[Composting/Landfills] Composting: Using organic
material to grow plants 1: 115

[Dissolved Oxygen] Decay and Dissolved Oxygen:
How does the amount of decaying matter affect
the level of dissolved oxygen in water? 1: 144

[Dissolved Oxygen] Goldfish Breath: How does a
decrease in the dissolved oxygen level affect
the breathing rate of goldfish? 1: 149

[Erosion] Erosion: Does soil type affect the amount
of water that runs off a hillside? 2: 233

[Erosion] Plants and Erosion: How do plants affect the
rate of soil erosion? 2: 239

[Greenhouse Effect] Creating a Greenhouse:
How much will the temperature rise inside
a greenhouse? 2: 294

[Greenhouse Effect] Fossil Fuels: What happens
when fossil fuels burn? 2: 300

[Groundwater Aquifers] Aquifers: How do they
become polluted? 2: 311

[Groundwater Aquifers] Groundwater: How can it
be cleaned? 2: 316

[Solar Energy] Capturing Solar Energy: Will seedlings
grow bigger in a greenhouse? 3: 579

[Solar Energy] Solar Cells: Will sunlight make a
motor run? 3: 584

[Rivers of Water] Weathering Erosion in Glaciers: How
does a river make a trench? 3: 717

[Rivers of Water] Stream Flow: Does the
stream meander? 3: 721

[Water Cycle] Temperature: How does temperature
affect the rate of evaporation? 3: 731

[Water Cycle] Surface Area: How does surface area
affect the rate of evaporation? 3: 736

Geology

[Earthquakes] Detecting an Earthquake: How can
movement of Earth's crust be measured? 1: 162

[Earthquakes] Earthquake Simulation: Is the destruction
greater at the epicenter? 1: 167

[Rocks and Minerals] Mineral Testing: What kind of
mineral is it? 3: 531

[Rocks and Minerals] Rock Classification: Is it igneous,
sedimentary or metamorphic? 3: 534

[Volcanoes] Model of a Volcano: Will it blow its top? 3: 687

[Volcanoes] Looking at a Seismograph: Can a volcanic
eruption be detected? 3: 691

Meteorology

[Weather] Wind: Measuring wind speed with a
homemade anemometer 3: 748

[Weather] Clouds: Will a drop in air temperature
cause a cloud to form? 3: 752

[Weather Forecasting] Dewpoint:
When will dew form? 3: 764

[Weather Forecasting] Air Pressure: How can air
pressure be measured? 3: 769

Physics

[Density and Buoyancy] Density: Can a scale of
relative density predict whether one material
floats on another? 1: 126

[Density and Buoyancy] Buoyancy: Does water
pressure affect buoyancy? 1: 132

[Electricity] Electrolytes: Do some solutions conduct
electricity better than others? 2: 188

[Electricity] Batteries: Can a series of homemade electric
cells form a "pile" strong enough to match the voltage
of a D-cell battery? 2: 193

[Electromagnetism] Magnetism: How can a magnetic
field be created and detected? 2: 206

[Electromagnetism] Electromagnetism: How can an
electromagnet be created? 2: 210

[Flight] Lift-Off: How can a glider be made to
fly higher? 2: 252

experiments by scientific field

[Flight] Helicopters, Propellers, and Centripetal Force:
Will it fly high? 2: 256
[Gravity] Gravity: How fast do different objects fall? 2: 280
[Gravity] Measuring Mass: How can a balance be made? 2: 285
[Heat] Conduction: Which solid materials are the best
conductors of heat? 2: 327
[Heat] Convection: How does heat move
through liquids? 2: 334
[Properties of Light] Looking for the Glow: Which objects
glow under black light? 2: 360
[Properties of Light] Refraction and Defraction:
Making a rainbow 2: 363
[Magnetism] Magnets: How do heat, cold, jarring,
and rubbing affect the magnetism of a nail? 3: 373
[Magnetism] Electromagnets: Does the strength of an
electromagnet increase with greater current? 3: 379
[Optics/Optical Illusions] Optics: What is the focal
length of a lens? 3: 432
[Optics/Optical Illusions] Optical Illusions: Can the
eye be fooled? 3: 437
[Potential and Kinetic Energy] Measuring Energy:
How does the height of an object affect its
potential energy? 3: 512
[Potential and Kinetic Energy] Using Energy: Build a
roller coaster 3: 517
[Solar Energy] Capturing Solar Energy: Will seedlings
grow bigger in a greenhouse? 3: 579
[Solar Energy] Solar Cells: Will sunlight make
a motor run? 3: 584
[Sound] Wave Length: How does the length of a vibrating
string affect the sound it produces? 3: 592
[Sound] Pitch: How does the thickness of a vibrating
string affect sound? 3: 596
[Static Electricity] Building an Electroscope:
Which objects are electrically charged? 3: 619
[Static Electricity] Measuring a Charge: Does nylon or
wool create a stronger static electric charge? 3: 625
[Structures and Shapes] Arches and Beams:
Which is strongest? 3: 637
[Structures and Shapes] Beams and Rigidity: How does
the vertical height of a beam affect its rigidity? 3: 640

All Subjects

[Scientific Method] Using the Scientific Method:
What are the mystery powders? 3: 562
[Scientific Method] Using the Scientific Method:
Do fruit flies appear out of thin air? 3: 568

**experiments
by scientific
field**

words to know

A

Abscission: The point at which a leaf meets a twig.

Acceleration: The rate at which the velocity and/or direction of an object is changing with the respect to time.

Acid: Substance that when dissolved in water is capable of reacting with a base to form salts and release hydrogen ions.

Acid rain: A form of precipitation that is significantly more acidic than neutral water, often produced as the result of industrial processes.

Acoustics: The science concerned with the production, properties, and propagation of sound waves.

Active solar energy system: A solar energy system that uses pumps or fans to circulate heat captured from the Sun.

Adhesion: Attraction between two different substances.

Aeration: Shaking a liquid to allow trapped gases to escape and to add oxygen.

Aerobic: Requiring oxygen.

Aerodynamics: The study of the motion of gases (particularly air) and the motion and control of objects in the air.

Alga/Algae: Single-celled or multicellular plants or plantlike organisms that contain chlorophyll, thus making their own food by photosynthesis. Algae grow mainly in water.

Alignment: Adjustment to a certain direction or orientation.

Alkaline: Having a pH of more than 7.

Alloy: A mixture of two or more metals with properties different from those metals of which it is made.

Amine: An organic compound derived from ammonia.

Amphibians: Animals that live on land and breathe air but return to the water to reproduce.

Amplitude: The maximum displacement (difference between an original position and a later position) of the material that is vibrating. Amplitude can be thought of visually as the highest and lowest points of a wave.

Anaerobic: Functioning without oxygen.

Anemometer: A device that measures wind speed.

Animalcules: Life forms that Anton van Leeuwenhoek named when he first saw them under his microscope; they later became known as protozoa and bacteria.

Anthocyanin: Red pigment found in leaves, petals, stems, and other parts of a plant.

Antibody: A protein produced by certain cells of the body as an immune (disease-fighting) response to a specific foreign antigen.

Aquifer: Underground layer of sand, gravel, or spongy rock that collects water.

Arch: A curved structure spanning an opening that supports a wall or other weight above the opening.

Artesian well: A well in which water is under pressure.

Asexual reproduction: Any reproductive process that does not involve the union of two individuals in the exchange of genetic material.

Astronomers: Scientists who study the positions, motions, and composition of stars and other objects in the sky.

Astronomy: The study of the physical properties of objects and matter outside Earth's atmosphere.

Atmosphere: Layers of air that surround Earth.

Atmospheric pressure: The pressure exerted by the atmosphere at Earth's surface due to the weight of the air.

Atom: The smallest unit of an element, made up of protons and neutrons in a central nucleus surrounded by moving electrons.

Autotroph: An organism that can build all the food and produce all the energy it needs with its own resources.

Auxins: A group of plant hormones responsible for patterns of plant growth.

B

Bacteria: Single-celled microorganisms that live in soil, water, plants, and animals that play a key role in the decaying of organic matter and the cycling of nutrients. Some are agents of disease.

Bacteriology: The scientific study of bacteria, their characteristics, and their activities as related to medicine, industry, and agriculture.

Base: Substance that when dissolved in water is capable of reacting with an acid to form salts and release hydrogen ions; has a pH of more than 7.

Beriberi: A disease caused by a deficiency of thiamine and characterized by nerve and gastrointestinal disorders.

Biochemical oxygen demand (BOD_5): The amount of oxygen that microorganisms use over a five-day period in 68° Fahrenheit (20° Celsius) water to decay organic matter.

Biological variables: Living factors such as bacteria, fungi, and animals that can affect the processes that occur in nature and in an experiment.

Biomes: Large geographical areas with specific climates and soils, as well as distinct plant and animal communities that are interdependent.

Bond: The force that holds two atoms together.

Botany: The branch of biology involving the study of plant life.

Braided rivers: Wide, shallow rivers with pebbly islands in the middle.

Buoyancy: The tendency of a fluid to exert a lifting effect on a body immersed in it.

By-products: Something produced in the making of something else.

c

Calibration: Standardizing or adjusting a measuring instrument so its measurements are correct.

Capillary action: The tendency of water to rise through a narrow tube by the force of adhesion between the water and the walls of the tube.

Carbohydrate: A compound consisting of carbon, hydrogen, and oxygen found in plants and used as a food by humans and other animals.

Carnivore: Meat-eating organism.

Carotene: Yellowish-orange pigment present in most leaves.

Catalyst: A compound that speeds up the rate of a chemical reaction without undergoing any change in its own composition.

Celestial: Describing planets or other objects in space.

Cell: The basic unit of a living organism; cells are structured to perform highly specialized functions.

Cell membrane: The thin layer of tissue that surrounds a cell.

Cell theory: The idea that all living things have one or more similar cells that carry out the same functions for the living process.

Centrifuge: A device that rapidly spins a solution so that the heavier components will separate from the lighter ones.

Centripetal force: Rotating force that moves towards the center or axis.

Channel: A shallow trench carved into the ground by the pressure and movement of a river.

Chemical energy: Energy stored in chemical bonds.

Chemical property: A characteristic of a substance that allows it to undergo a chemical change. Chemical properties include flammability and sensitivity to light.

Chemical reaction: Any chemical change in which at least one new substance is formed.

Chlorophyll: A green pigment found in plants that absorbs sunlight, providing the energy used in photosynthesis, or the conversion of carbon dioxide and water to complex carbohydrates.

Chloroplasts: Small structures in plant cells that contain chlorophyll and in which the process of photosynthesis takes place.

Chromatography: A method for separating mixtures into their component parts (into their "ingredients," or into what makes them up).

Circuit: The complete path of an electric current including the source of electric energy.

Cleavage: The tendency of a mineral to split along certain planes.

Climate: The average weather that a region experiences over a long period.

Coagulation: The clumping together of particles in a liquid.

Cohesion: Attraction between like substances.

Colloid: A mixture containing particles suspended in, but not dissolved in, a dispersing medium.

Colony: A mass of microorganisms that have been bred in a medium.

Combustion: Any chemical reaction in which heat, and usually light, is produced. It is commonly the burning of organic substances during which oxygen from the air is used to form carbon dioxide and water vapor.

Complete metamorphosis: Metamorphosis in which a larva becomes a pupa before changing into an adult form.

Composting: The process in which organic compounds break down and become dark, fertile soil called humus.

Concave: Hollowed or rounded upward, like the inside of a bowl; arched.

Concentration: The amount of a substance present in a given volume, such as the number of molecules in a liter.

Condense/condensation: The process by which a gas changes into a liquid.

Conduction: The flow of heat through a solid.

Confined aquifer: An aquifer with a layer of impermeable rock above it; the water is held under pressure.

Coniferous: Refers to trees, such as pines and firs, that bear cones and have needle-like leaves that are not shed all at once.

Constellations: Eighty-eight patterns of stars in the night sky.

Continental drift: The theory that continents move apart slowly at a predictable rate.

Control experiment: A set-up that is identical to the experiment but is not affected by the variable that will be changed during the experiment.

Convection: The circulatory motion that occurs in a gas or liquid at a nonuniform temperature; the variation of the motion is caused by the substance's density and the action of gravity.

Convection current: Circular movement of a fluid in response to alternating heating and cooling.

Convex: Curved or rounded like the outside of a ball.

Corona: The outermost atmospheric layer of the Sun.

Corrosion: An oxidation-reduction reaction in which a metal is oxidized (reacted with oxygen) and oxygen is reduced, usually in the presence of moisture.

Cotyledon: Seed leaves, which contain stored food for the embryo.

Crust: The hard, outer shell of Earth that floats upon the softer, denser mantle.

Cultures: Microorganisms growing in prepared nutrients.

Cyanobacteria: Oxygen-producing, aquatic bacteria capable of manufacturing its own food; resembles algae.

Cycle: Occurrence of events that take place the same time every year; a single complete vibration.

Cytology: The branch of biology concerned with the study of cells.

Cytoplasm: The semifluid substance inside a cell that surrounds the nucleus and the other membrane-enclosed organelles.

D

Decanting: The process of separating a suspension by waiting for its heavier components to settle out and then pouring off the lighter ones.

Decibel (dB): A unit of measurement for sound.

Deciduous: Plants that lose their leaves at some season of the year, and then grow them back at another season.

Decomposition: The breakdown of complex molecules—molecules of which dead organisms are composed—into simple nutrients that can be reutilized by living organisms.

Decomposition reaction: A chemical reaction in which one substance is broken down into two or more substances.

Denaturization: Altering of an enzyme so it no longer works.

Density: The mass of a substance compared to its volume.

Density ball: A ball with the fixed standard of 1.0 g/l, which is the exact density of pure water.

Dependent variable: The variable in a function whose value depends on the value of another variable in the function.

Deposition: Dropping of sediments that occurs when a river loses its energy of motion.

Desert: A biome with a hot-to-cool climate and dry weather.

Desertification: Transformation of arid or semiarid productive land into desert.

Dewpoint: The point at which water vapor begins to condense.

Dicot: Plants with a pair of embryonic seeds that appear at germination.

Diffraction: The bending of light or another form of electromagnetic radiation as it passes through a tiny hole or around a sharp edge.

Diffraction grating: A device consisting of a surface into which are etched very fine, closely spaced grooves that cause different wavelengths of light to reflect or refract (bend) by different amounts.

Diffusion: Random movement of molecules that leads to a net movement of molecules from a region of high concentration to a region of low concentration.

Disinfection: Using chemicals to kill harmful organisms.

Dissolved oxygen (DO): Oxygen molecules that have dissolved in water.

Distillation: The process of separating liquids from solids or from other liquids with different boiling points by a method of evaporation and condensation, so that each component in a mixture can be collected separately in its pure form.

DNA: Abbreviation for deoxyribonucleic acid. Large, complex molecules found in nuclei of cells that carry genetic information for an organism's development.

Domain: Small regions in an iron object that possess their own magnetic charges.

Dormancy: A state of inactivity in an organism.

Dormant: Describing an inactive organism.

Drought: A prolonged period of dry weather that damages crops or prevents their growth.

Dry cell: An electrolytic cell or battery using a non-liquid electrolyte.

Dynamic equilibrium: A situation in which substances are moving into and out of cell walls at an equal rate.

E

Earthquake: An unpredictable event in which masses of rock shift below Earth's surface, releasing enormous amounts of energy and sending out shock waves that sometimes cause the ground to shake dramatically.

Eclipse: A phenomenon in which the light from a celestial body is temporarily cut off by the presence of another body.

Ecologists: Scientists who study the interrelationship of organisms and their environments.

Ecosystem: An ecological community, including plants, animals and microorganisms considered together with their environment.

Electric charge repulsion: Repulsion of particles caused by a layer of negative ions surrounding each particle. The repulsion prevents coagulation and promotes the even dispersion of such particles through a mixture.

Electrical energy: The motion of electrons within any object that conducts electricity.

Electricity: A form of energy caused by the presence of electrical charges in matter.

Electrode: A material that will conduct an electrical current, usually a metal; used to carry electrons into or out of an electrochemical cell.

Electrolyte: Any substance that, when dissolved in water, conducts an electric current.

Electromagnetic spectrum: The complete array of electromagnetic radiation, including radio waves (at the longest-wavelength end), microwaves, infrared radiation, visible light, ultraviolet radiation, X rays, and gamma rays (at the shortest-wavelength end).

Electromagnetic waves: Radiation that has properties of both an electric and a magnetic wave and that travels through a vacuum at the speed of light.

Electromagnetism: A form of magnetic energy produced by the flow of an electric current through a metal core. Also, the study of electric and magnetic fields and their interaction with charges and currents.

Electron: A subatomic particle with a mass of about one atomic mass unit and a single electrical charge that orbits the nucleus of an atom.

Electroscope: A device that determines whether an object is electrically charged.

Elevation: Height above sea level.

Elliptical: An orbital path that is egg-shaped or resembles an elongated circle.

Embryo: The seed of a plant, which through germination can develop into a new plant; also, the earliest stage of animal development.

Embryonic: The earliest stages of development.

Endothermic reaction: A chemical reaction that absorbs energy, such as photosynthesis, the production of food by plant cells.

Energy: The ability to cause an action or for work to be done. Also, power that can be used to perform work, such as solar energy.

Environmental variables: Nonliving factors such as air temperature, water, pollution, and pH that can affect processes that occur in nature and in an experiment.

Enzymes: Any of numerous complex proteins produced by living cells that act as catalysts, speeding up the rate of chemical reactions in living organisms.

Enzymology: The science of studying enzymes.

Ephemerals: Plants that lie dormant in dry soil for years until major rainstorms occur.

Epicenter: The location where the seismic waves of an earthquake first appear on the surface, usually almost directly above the focus.

Equilibrium: A process in which the rates at which various changes take place balance each other, resulting in no overall change.

Erosion: The process by which topsoil is carried away by water, wind, or ice.

Eutrophic zone: The upper part of the ocean where sunlight penetrates, supporting plant life such as phytoplankton.

Eutrophication: Natural process by which a lake or other body of water becomes enriched in dissolved nutrients, spurring aquatic plant growth.

Evaporate/evaporation: The process by which liquid changes into a gas; also, the escape of water vapor into the air, yielding only the solute.

Exothermic reaction: A chemical reaction that releases energy, such as the burning of fuel.

Experiment: A controlled observation.

F

Fat: A type of lipid, or chemical compound used as a source of energy, to provide insulation, and to protect organs in an animal's body.

Fault: A crack running through rock that is the result of tectonic forces.

Fault blocks: Pieces of rock from Earth's crust that overlap and cause earthquakes when they press together and snap from pressure.

Filtration: The use of a screen or filter to separate larger particles from smaller ones that can slip through the filter's openings.

Fluorescence: Luminescence (glowing) that stops within 10 nanoseconds after an energy source has been removed.

Focal length: The distance of a focus from the surface of a lens or concave mirror.

Focal point: The point at which rays of light converge (come together) or from which they diverge (move apart).

Food web: An interconnected set of all the food chains in the same ecosystem.

Force: A physical interaction (pushing or pulling) tending to change the state of motion (velocity) of an object.

Fossil fuel: A fuel such as coal, oil, or natural gas that is formed over millions of years from the remains of plants and animals.

Fracture: A mineral's tendency to break into curved, rough, or jagged surfaces.

Frequency: The rate at which vibrations take place (number of times per second the motion is repeated), given in cycles per second or in hertz (Hz). Also, the number of waves that pass a given point in a given period of time.

Front: The front edges of moving masses of air.

Fungus: Kingdom of various single-celled or multicellular organisms, including mushrooms, molds, yeasts, and mildews, that do not contain chlorophyll. (Plural is fungi.)

Fusion: Combining of nuclei of two or more lighter elements into one nucleus of a heavier element; the process stars use to produce energy to support themselves against their own gravity.

G

Galaxy: A large collection of stars and clusters of stars containing anywhere from a few million to a few trillion stars.

Gene: A segment of a DNA (deoxyribonucleic acid) molecule contained in the nucleus of a cell that acts as a kind of code for the production of some specific protein. Genes carry instructions for the formation, functioning, and transmission of specific traits from one generation to another.

Genetic material: Material that transfers characteristics from a parent to its offspring.

Geology: The study of the origin, history, and structure of Earth.

Geotropism: The tendency of roots to bend toward Earth.

Germ theory of disease: The belief that disease is caused by germs.

Germination: The beginning of growth of a seed.

Gibbous moon: A phase of the Moon when more than half of its surface is lighted.

Glacier: A large mass of ice formed from snow that has packed together and which moves slowly down a slope under its own weight.

Global warming: Warming of Earth's atmosphere that results from an increase in the concentration of gases that store heat such as carbon dioxide.

Glucose: Also known as blood sugar; a simple sugar broken down in cells to produce energy.

Golgi body: Organelle that sorts, modifies, and packages molecules.

Gravity: Force of attraction between objects, the strength of which depends on the mass of each object and the distance between them.

Greenhouse effect: The warming of Earth's atmosphere due to water vapor, carbon dioxide, and other gases in the atmosphere that trap heat radiated from Earth's surface.

Greenhouse gases: Gases that absorb infrared radiation and warm air before it escapes into space.

Groundwater: Water that soaks into the ground and is stored in the small spaces between the rocks and soil.

H

Heat: A form of energy produced by the motion of molecules that make up a substance.

Heat energy: The energy produced when two substances that have different temperatures are combined.

Herbivore: Plant-eating organism.

Hertz (Hz): The unit of frequency; a measure of the number of waves that pass a given point per second of time.

Heterotrophs: Organisms that cannot make their own food and that must, therefore, obtain their food from other organisms.

High air pressure: An area where the air molecules are more dense.

Hormone: A chemical produced in living cells that regulates the functions of the organism.

Humidity: The amount of water vapor (moisture) contained in the air.

Humus: Fragrant, spongy, nutrient-rich decayed plant or animal matter.

Hydrologic cycle: Continual movement of water from the atmosphere to Earth's surface through precipitation and back to the atmosphere through evaporation and transpiration.

Hydrologists: Scientists who study water and its cycle.

Hydrology: The study of water and its cycle.

Hydrometer: An instrument that determines the specific gravity of a liquid.

Hydrophilic: A substance that is attracted to and readily mixes with water.

Hydrophobic: A substance that is repelled by and does not mix with water.

Hydrotropism: The tendency of roots to grow toward a water source.

Hypertonic solution: A solution with a higher osmotic pressure (solute concentration) than another solution.

Hypothesis: An idea in the form of a statement that can be tested by observation and/or experiment.

Hypotonic solution: A solution with a lower osmotic pressure (solute concentration) than another solution.

I

Igneous rock: Rock formed from the cooling and hardening of magma.

Immiscible: Incapable of being mixed.

Impermeable: Not allowing substances to pass through.

Impurities: Chemicals or other pollutants in water.

Incomplete metamorphosis: Metamorphosis in which a nymph form gradually becomes an adult through molting.

Independent variable: The variable in a function that determines the final value of the function.

Indicator: Pigments that change color when they come into contact with acidic or basic solutions.

Inertia: The tendency of an object to continue in its state of motion.

Infrared radiation: Electromagnetic radiation of a wavelength shorter than radio waves but longer than visible light that takes the form of heat.

Inner core: Very dense, solid center of Earth.

Inorganic: Not made of or coming from living things.

Insulated wire: Electrical wire coated with a nonconducting material such as plastic.

Insulation/insulator: A material that does not conduct heat or electricity.

Interference fringes: Bands of color that fan around an object.

Ion: An atom or group of atoms that carries an electrical charge—either positive or negative—as a result of losing or gaining one or more electrons.

Ionic conduction: The flow of an electrical current by the movement of charged particles, or ions.

Isobars: Continuous lines on a map that connect areas with the same air pressure.

Isotonic solutions: Two solutions that have the same concentration of solute particles and therefore the same osmotic pressure.

K

Kinetic energy: Energy of an object or system due to its motion.

L

Lactobacilli: A strain of bacteria.

Larva: Immature form (wormlike in insects; fishlike in amphibians) of an organism capable of surviving on its own. A larva does not resemble the parent and must go through metamorphosis, or change, to reach its adult stage.

Lava: Molten rock that occurs at the surface of Earth, usually through volcanic eruptions.

Lens: A piece of transparent material with two curved surfaces that bring together and focus rays of light passing through it.

Lichen: An organism composed of a fungus and a photosynthetic organism in a symbiotic relationship.

Lift: Upper force on the wings of an aircraft created by differences in air pressure on top of and underneath the wings.

Light-year: Distance light travels in one year in the vacuum of space, roughly 5.9 trillion miles (9.5 trillion km).

The Local Group: A cluster of 30 galaxies, including the Milky Way, pulled together gravitationally.

Low air pressure: An area where the air molecules are less dense.

Lunar eclipse: Eclipse that occurs when Earth passes between the Sun and the Moon, casting a shadow on the Moon.

Luster: A glow of reflected light; a sheen.

M

Macroorganisms: Visible organisms that aid in breaking down organic matter.

Magma: Molten rock deep within Earth that consists of liquids, gases, and particles of rocks and crystals. Magma underlies areas of volcanic activity and at Earth's surface is called lava.

Magma chambers: Pools of bubbling liquid rock that are the energy sources causing volcanoes to be active.

Magma surge: A swell or rising wave of magma caused by the movement and friction of tectonic plates; the surge heats and melts rock, adding to the magma and its force.

Magnet: A material that attracts other like material, especially metals.

Magnetic circuit: A series of magnetic domains aligned in the same direction.

Magnetic field: The space around an electric current or a magnet in which a magnetic force can be observed.

Magnetism: A fundamental force of nature caused by the motion of electrons in an atom. Magnetism is manifested by the attraction of certain materials for iron.

Mantle: Thick, dense layer of rock that underlies Earth's crust and overlies the core.

Manure: The waste matter of animals.

Mass: Measure of the total amount of matter in an object. Also, an object's quantity of matter as shown by its gravitational pull on another object.

Matter: Anything that has mass and takes up space.

Meandering river: A lowland river that twists and turns along its route to the sea.

Medium: A material that carries the acoustic vibrations away from the body producing them.

Meniscus: The curved surface of a column of liquid.

Metamorphic rock: Rock formed by transformation of pre-existing rock through changes in temperature and pressure.

Metamorphosis: Transformation of an immature animal into an adult.

Meteorologists: Scientists who study weather and weather forecasting.

Microbiology: Branch of biology dealing with microscopic forms of life.

Microclimate: A local climate.

Microorganisms: Living organisms so small that they can be seen only with the aid of a microscope.

Micropyle: Seed opening that enables water to enter easily.

Milky Way: The galaxy in which our solar system is located.

Mineral: An inorganic substance found in nature with a definite chemical composition and structure. As a nutrient, helps build bones and soft tissues and regulates body functions.

Mixtures: Combinations of two or more substances that are not chemically combined with each other and can exist in any proportion.

Molecule: The smallest particle of a substance that retains all the properties of the substance and is composed of one or more atoms.

Molting: Shedding of the outer layer of an animal, as occurs during growth of insect larvae.

Monocot: Plants with a single embryonic seed at germination.

Moraine: Mass of boulders, stones, and other rock debris carried along and deposited by a glacier.

Multicellular: Living things with many cells joined together.

N

Nanometer: A unit of length; this measurement is equal to one-billionth of a meter.

Nansen bottles: Self-closing containers with thermometers that draw in water at different depths.

Nebula: Bright or dark cloud, often composed of gases and dust, hovering in the space between the stars.

Neutralization: A chemical process in which the mixing of an acidic solution with a basic (alkaline) solution results in a solution that has the properties of neither an acid nor a base.

Neutron: A subatomic particle with a mass of about one atomic mass unit and no electrical charge that is found in the nucleus of an atom.

Niche: The specific role that an organism carries out in its ecosystem.

Nonpoint source: An unidentified source of pollution; may actually be a number of sources.

Nucleus: The central core of an atom, consisting of protons and (usually) neutrons.

Nutrient: A substance needed by an organism in order for it to survive, grow, and develop.

Nutrition: The study of the food nutrients an organism needs in order to maintain well-being.

Nymph: An immature form in the life cycle of insects that go through an incomplete metamorphosis.

o

Oceanography: The study of the chemistry of the oceans, as well as their currents, marine life, and the ocean bed.

Optics: The study of the nature of light and its properties.

Organelles: Membrane-bounded cellular "organs" performing a specific set of functions within a eukaryotic cell.

Organic: Made of or coming from living things.

Osmosis: The movement of fluids and substances dissolved in liquids across a semipermeable membrane from an area of its greater concentration to an area of its lesser concentration until all substances involved reach a balance.

Outer core: A liquid core that surrounds Earth's solid inner core; made mostly of iron.

Oxidation: A chemical reaction in which oxygen reacts with some other substance and in which ions, atoms, or molecules lose electrons.

Oxidation-reduction reaction: A chemical reaction in which one substance loses one or more electrons and the other substance gains one or more electrons.

Oxidation state: The sum of an atom's positive and negative charges.

Oxidizing agent: A chemical substance that gives up oxygen or takes on electrons from another substance.

Ozone layer: The atmospheric layer of approximately 15 to 30 miles (24 to 48 km) above Earth's surface in which the concentration of ozone is significantly higher than in other parts of the atmosphere and that protects the lower atmosphere from harmful solar radiation.

P

Papain: An enzyme obtained from the fruit of the papaya used as a meat tenderizer, as a drug to clean cuts and wounds, and as a digestive aid for stomach disorders.

Passive solar energy system: A solar energy system in which the heat of the Sun is captured, used, and stored by means of the design of a building and the materials from which it is made.

Pasteurization: The process of slow heating that kills bacteria and other microorganisms.

Penicillin: A mold from the fungi group of microorganisms used as an antibiotic.

Pepsin: Digestive enzyme that breaks down protein.

Percolate: To pass through a permeable substance.

Permeable: Having pores that permit a liquid or a gas to pass through.

pH: Abbreviation for potential hydrogen. A measure of the acidity or alkalinity of a solution determined by the concentration of hydrogen ions present in a liter of a given fluid. The pH scale ranges from 0 (greatest concentration of hydrogen ions and therefore most acidic) to 14 (least concentration of hydrogen ions and therefore most alkaline), with 7 representing a neutral solution, such as pure water.

Pharmacology: The science dealing with the properties, reactions, and therapeutic values of drugs.

Phases: Changes in the illuminated Moon surfaces as the Moon revolves around Earth.

Phloem: Plant tissue consisting of elongated cells that transport carbohydrates and other nutrients.

Phosphorescence: Luminescence (glowing) that stops within 10 nanoseconds after an energy source has been removed.

Photoelectric effect: The phenomenon in which light falling upon certain metals stimulates the emission of electrons and changes light into electricity.

Photosynthesis: Chemical process by which plants containing chlorophyll use sunlight to manufacture their own food by converting carbon dioxide and water to carbohydrates, releasing oxygen as a by-product.

Phototropism: The tendency of a plant to grow toward a source of light.

Photovoltaic cells: A device made of silicon that converts sunlight into electricity.

Physical change: A change in which the substance keeps its identity, such as a piece of chalk that has been ground up.

Physical property: A characteristic that you can detect with your senses, such as color and shape.

Phytoplankton: Microscopic aquatic plants that live suspended in the water.

Pigment: A substance that displays a color because of the wavelengths of light that it reflects.

Pitch: A property of a sound, determined by its frequency; the highness or lowness of a sound.

Plates: Large regions of Earth's surface, composed of the crust and uppermost mantle, which move about, forming many of Earth's major geologic surface features.

Pnematocysts: Stinging cells.

Point source: An identified source of pollution.

Pollination: The transfer of pollen from the male reproductive organs to the female reproductive organs of plants.

Pore: An opening or space.

Potential energy: The energy possessed by a body as a result of its position.

Precipitation: Water in its liquid or frozen form when it falls from clouds as rain, snow, sleet, or hail.

Probe: The terminal of a voltmeter, used to connect the voltmeter to a circuit.

Producer: An organism that can manufacture its own food from nonliving materials and an external energy source, usually by photosynthesis.

experiment
CENTRAL

Product: A compound that is formed as a result of a chemical reaction.

Prominences: Masses of glowing gas, mainly hydrogen, that rise from the Sun's surface like flames.

Propeller: Radiating blades mounted on a quickly rotating shaft that are used to move aircraft forward.

Protein: A complex chemical compound that consists of many amino acids attached to each other that are essential to the structure and functioning of all living cells.

Protists: Members of the kingdom Protista, primarily single-celled organisms that are not plants or animals.

Proton: A subatomic particle with a mass of about one atomic mass unit and a single negative electrical change that is found in the nucleus of an atom.

Protozoan: Single-celled animal-like microscopic organisms that live by taking in food rather than making it by photosynthesis and must live in the presence of water. (Plural is protozoa.)

Pupa: A stage in the metamorphosis of an insect during which its tissues are completely reorganized to take on their adult shape.

R

Radiation: Energy transmitted in the form of electromagnetic waves or subatomic particles.

Radicule: A seed's root system.

Radio wave: Longest form of electromagnetic radiation, measuring up to 6 miles (9.6 km) from peak to peak.

Radiosonde balloons: Instruments for collecting data in the atmosphere and then transmitting that data back to Earth by means of radio waves.

Reactant: A compound present at the beginning of a chemical reaction.

Reaction: Response to an action prompted by a stimulus.

Reduction: A process in which a chemical substance gives off oxygen or takes on electrons.

Reflection: The bouncing of light rays in a regular pattern off the surface of an object.

Refraction: The bending of light rays as they pass at an angle from one transparent or clear medium into a second one of different density.

Rennin: Enzyme used in making cheese.

Resistance: A partial or complete limiting of the flow of electrical current through a material.

Respiration: The physical process that supplies oxygen to living cells and the chemical reactions that take place inside the cells.

Resultant: A force that results from the combined action of two other forces.

Retina: The light-sensitive part of the eyeball that receives images and transmits visual impulses through the optic nerve to the brain.

River: A main course of water into which many other smaller bodies of water flow.

Rock: Naturally occurring solid mixture of minerals.

Runoff: Water in excess of what can be absorbed by the ground.

S

Salinity: The amount of salts dissolved in seawater.

Saturated: Containing the maximum amount of a solute for a given amount of solvent at a certain temperature.

Scientific method: Collecting evidence meticulously and then theorizing from it.

Scribes: Ancient scholars.

Scurvy: A disease caused by a deficiency of vitamin C, which causes a weakening of connective tissue in bone and muscle.

Sediment: Sand, silt, clay, rock, gravel, mud, or other matter that has been transported by flowing water.

Sedimentary rock: Rock formed from the compressed and solidified layers of organic or inorganic matter.

Sedimentation: A process during which gravity pulls particles out of a liquid.

Seismic belt: Boundaries where Earth's plates meet.

Seismic waves: Classified as body waves or surface waves, vibrations in rock and soil that transfer the force of the earthquake from the focus (center) into the surrounding area.

Seismograph: A device that records vibrations of the ground and within Earth.

Seismology: The study and measurement of earthquakes.

Seismometer: A seismograph that measures the movement of the ground.

Semipermeable membrane: A thin barrier between two solutions that permits only certain components of the solutions, usually the solvent, to pass through.

Sexual reproduction: A reproductive process that involves the union of two individuals in the exchange of genetic material.

Silt: Medium-sized soil particles.

Solar collector: A device that absorbs sunlight and collects solar heat.

Solar eclipse: Eclipse that occurs when the Moon passes between Earth and the Sun, casting a shadow on Earth.

Solar energy: Any form of electromagnetic radiation that is emitted by the Sun.

Solute: The substance that is dissolved to make a solution and exists in the least amount in a solution, for example sugar in sugar water.

Solution: A mixture of two or more substances that appears to be uniform throughout except on a molecular level.

Solvent: The major component of a solution or the liquid in which some other component is dissolved, for example water in sugar water.

Specific gravity: The ratio of the density of a substance to the density of another substance.

Spectrum: Range of individual wavelengths of radiation produced when white light is broken down into its component colors when it passes through a prism or is broken apart by some other means.

Standard: A base for comparison.

Star: A vast clump of hydrogen gas and dust that produces great energy through fusion reactions at its core.

Static electricity: A form of electricity produced by friction in which the electric charge does not flow in a current but stays in one place.

Streak: The color of the dust left when a mineral is rubbed across a surface.

Substrate: The substance on which an enzyme operates in a chemical reaction.

Succulent: Plants that live in dry environments and have water storage tissue.

Surface water: Water in lakes, rivers, ponds, and streams.

Suspension: A temporary mixture of a solid in a gas or liquid from which the solid will eventually settle out.

Symbiosis: A pattern in which two or more organisms live in close connection with each other, often to the benefit of both or all organisms.

Synthesis reaction: A chemical reaction in which two or more substances combine to form a new substance.

T

Taiga: A large land biome mostly dominated by coniferous trees.

Tectonic plates: Huge flat rocks that form Earth's crust.

Temperate: Mild or moderate weather conditions.

Temperature: The measure of the average energy of the molecules in a substance.

Terminal: A connection in an electric circuit; usually a connection on a source of electric energy such as a battery.

Terracing: A series of horizontal ridges made in a hillside to reduce erosion.

Testa: A tough outer layer that protects the embryo and endosperm of a seed from damage.

Thermal conductivity: A number representing a material's ability to conduct heat.

Thermal energy: Energy caused by the movement of molecules due to the transfer of heat.

Thiamine: A vitamin of the B complex that is essential to normal metabolism and nerve function.

Thigmotropism: The tendency for a plant to grow toward a surface it touches.

Titration: A procedure in which an acid and a base are slowly mixed to achieve a neutral substance.

Toxic: Poisonous.

Trace element: A chemical element present in minute quantities.

Translucent: Permits the passage of light.

Tropism: The growth or movement of a plant toward or away from a stimulus.

Troposphere: The lowest layer of Earth's atmosphere, ranging to an altitude of about 9 miles (15 km) above Earth's surface.

Tsunami: A tidal wave caused by an earthquake.

Tuber: An underground, starch-storing stem, such as a potato.

Tundra: A treeless, frozen biome with low-lying plants.

Turbulence: Air disturbance or unrest that affects an aircraft's flight.

Tyndall effect: The effect achieved when colloidal particles reflect a beam of light, making it visible when shined through such a mixture.

U

Ultraviolet: Electromagnetic radiation (energy) of a wavelength just shorter than the violet (shortest wavelength) end of the visible light spectrum and thus with higher energy than the visible light.

Unconfined aquifer: An aquifer under a layer of permeable rock and soil.

Unicellular: Living things that have one cell. Protozoans are unicellular.

Universal gravitation: The notion of the constancy of the force of gravity between two bodies.

V

Vacuole: A space-filling organelle of plant cells.

Variable: Something that can change the results of an experiment.

Vegetative propagation: A form of asexual reproduction in which plants are produced that are genetically identical to the parent.

Viable: The capability of developing or growing under favorable conditions.

Vibration: A regular, back-and-forth motion of molecules in the air.

Visible spectrum: Light waves visible to the eye.

Vitamin: A complex organic compound found naturally in plants and animals that the body needs in small amounts for normal growth and activity.

Volcano: A conical mountain or dome of lava, ash, and cinders that forms around a vent leading to molten rock deep within Earth.

Voltage: Also called potential difference; the amount of electric energy stored in a mass of electric charges compared to the energy stored in some other mass of charges.

Voltmeter: An instrument for measuring the conductivity or resistance in a circuit or the voltage produced by an electric source.

Volume: The amount of space occupied by a three-dimensional object; the amplitude or loudness of a sound.

W

Water (hydrologic) cycle: The constant movement of water molecules on Earth as they rise into the atmosphere as water vapor, condense into droplets and fall to land or bodies of water, evaporate, and rise again.

Waterline: The highest point to which water rises on the hull of a ship. The portion of the hull below the waterline is under water.

Water table: The upper surface of groundwater.

Water vapor: Water in its gaseous state.

Wave: A motion in which energy and momentum is carried away from some source.

Wavelength: The distance between the peak of a wave of light, heat, or energy and the next corresponding peak.

Weather: The state of the troposphere at a particular time and place.

Weather forecasting: The scientific predictions of future weather patterns.

Weight: The gravitational attraction of Earth on an object; the measure of the heaviness of an object.

Wetlands: Areas that are wet or covered with water for at least part of the year.

X

Xanthophyll: Yellow pigment in plants.

Xerophytes: Plants that require little water to survive.

Xylem: Plant tissue consisting of elongated, thick-walled cells that transport water and mineral nutrients.

experiment
CENTRAL

chapter 1

Acid Rain

Did you know that acid rain can also be acid snow, acid fog, or even acid dust? **Acid rain** is a form of precipitation that is significantly more acidic than neutral water. The pH scale offers a way to compare the acidity of substances, including rain. **pH** (the abbreviation for potential hydrogen) is a measure of the acidity or alkalinity of a solution. The symbol pH refers to the concentration of hydrogen **ions** present in a liter of fluid. The pH scale ranges from 0 (greatest concentration of hydrogen ions and therefore most acidic) to 14 (least concentration of hydrogen ions and therefore most **alkaline**). An alkaline solution is also called a **base**. The number 7 represents a neutral solution, such as pure water.

Water with a pH of 4 is ten times more acidic than water with a pH of 5. A pH of 4 is one hundred times more acidic than a pH of 6. So you can see that a small increase or decrease in pH makes a big difference in acid levels.

How does acid get in rain?

Normal rainfall is slightly acidic, with a pH of about 5.6. Rain with a pH below 5.6 is considered to be acid rain. Acid rain is created when smoke and fumes from burning **fossil fuels**—coal, oil, and natural gas—rise into the air. The smoke and fumes come from oil- and coal-fired power plants, factory smokestacks, and automobile exhaust.

The main **toxic** (poisonous) chemicals in this pollution are sulfur dioxide and nitrogen oxides. These chemicals react with sunlight and moisture in the air to produce rain or snow that is a mild solution of

Words to Know

Acid rain:
A form of precipitation that is significantly more acidic than neutral water, often produced as the result of industrial processes and pollution.

Alkaline:
Having a pH of more than 7.

Amphibians:
Animals that live both on land and in water.

Base:
A water-soluble compound that when dissolved in water makes an alkaline, or basic, solution with a pH of more than 7.

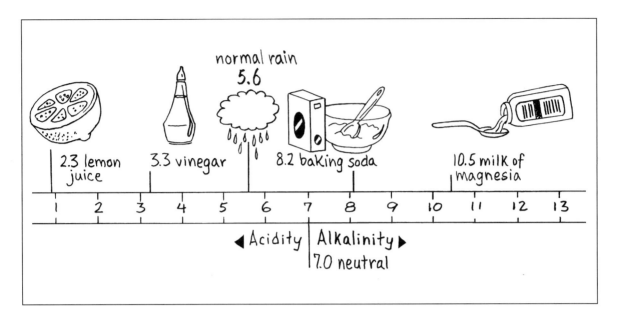

normal rain
5.6

2.3 lemon juice

3.3 vinegar

8.2 baking soda

10.5 milk of magnesia

| 1 | 2 | 3 | 4 | 5 | 6 | 7 | 8 | 9 | 10 | 11 | 12 | 13 |

◀ Acidity | Alkalinity ▶

7.0 neutral

The pH scale shows the acidity and alkalinity of liquids.

sulfuric acid and nitric acid. Some of the pollutant particles fall to the ground as acid dust. When acid rain falls, this dust dissolves in the water, further increasing the rain's acidity.

Why is acid rain a problem?

Acid rain can make lakes and streams so toxic that nothing can live there. **Amphibians** and the young of most fish are sensitive to acidity, so they are the first to die. With water at a pH of 5.0, most fish eggs are unable to hatch. If the pH level continues to drop, adult animals begin to die. Experiment 1 will help you determine how sensitive brine shrimp are to acid rain.

Acidity kills plants in the water, too, thus upsetting the food chain. Even plant-eating fish that can tolerate low pH levels are soon unable to find enough to eat. With few plant-eating fish able to survive, the fish-eating fish go hungry, too.

Acid rain can slowly kill whole forests by dissolving the toxic metals in soil and rock. In their dissolved form, these metals damage tree roots. Acid rain also dissolves nutrients in the soil and washes them away before the trees and plants can use them. In addition, acid rain burns tree leaves and needles and wears away their protective coatings, leaving them unable to produce enough food energy to meet the trees' needs. Viruses, fungi, and pests can then easily finish off the weakened trees. Experiment 2 will help you determine how acid rain affects plant growth.

Words to Know

Control experiment:
A set-up that is identical to the experiment but is not affected by the variable that will be changed during the experiment.

Fossil fuel:
A fuel such as coal, oil, or natural gas that is formed over millions of years from the remains of plants and animals.

Hypothesis:
An idea in the form of a statement that can be tested by observation and/or experiment.

(W)ords to Know

Ion:
An atom or groups of atoms that carry an electrical charge—either positive or negative—as a result of losing or gaining one or more electrons.

Neutralization:
A chemical process in which the mixing of an acidic solution with a basic (alkaline) solution results in a solution that has the properties of neither an acid nor a base.

pH:
A measure of the acidity or alkalinity of a solution referring to the concentration of hydrogen ions present in a liter of a given fluid. The pH scale ranges from 0 (greatest concentration of hydrogen ions and therefore most acidic) to 14 (least concentration of hydrogen ions and therefore most alkaline), with 7 representing a neutral solution, such as pure water.

What can be done?

Acid rain was first identified in 1852 by an English chemist named Robert Angus Smith. He suggested that factories that burned coal were sending sulfur dioxide into the air. Since then, the world has gained many more factories—and many more sources of air pollution.

Fortunately, scientists have found ways to wash the sulfur out of coal before it is burned and to wash the sulfur out of smoke before it leaves the smokestacks. In addition, new vehicles must now have a device called a catalytic converter, which uses filters and chemicals to change carbon monoxide and other air pollutants into carbon dioxide and water. This device nearly eliminates the nitrogen oxide released by cars' exhaust systems.

Lime, which is a natural base, can be added to streams and lakes to neutralize their acidity. **Neutralization** is a chemical process in which an acidic solution is mixed with a basic (alkaline) solution,

Trees take a long time to recover from damage caused by acid rain.

resulting in a solution that is neutral—it has the properties of neither an acid nor a base. However, neutralizing streams and lakes is expensive and must continue as long as acid rain keeps falling.

Scientists are also researching more ways to use sources of energy that do not pollute the air, including solar power. We all can help reduce acid rain by reducing our own use of fossil fuels and by learning more about the effects of acid rain.

Experiment 1
Acid Rain and Animals: How does acid rain affect brine shrimp?

Purpose/Hypothesis
In this experiment, you will use vinegar, which is an acid, to gradually lower the pH level of water containing brine shrimp. (As the pH level drops, acidity increases.) You will measure the changing pH level and observe how the shrimp react.

Before you begin, make an educated guess about the outcome of this experiment based on your knowledge of acid rain. This educated guess, or prediction, is your **hypothesis**. A hypothesis should explain these things:

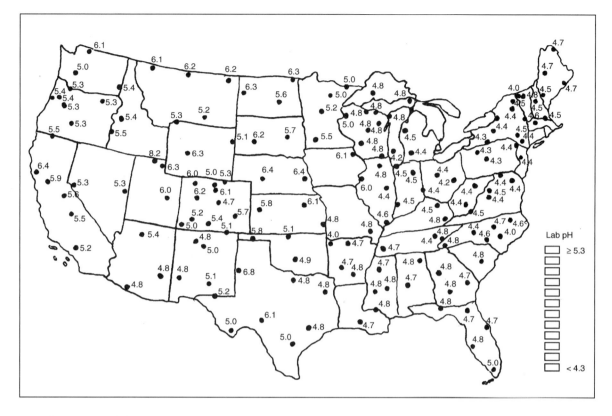

pH levels in the
United States.

- the topic of the experiment
- the variable you will change
- the variable you will measure
- what you expect to happen

A hypothesis should be brief, specific, and measurable. It must be something you can test through observation. Your experiment will prove or disprove whether your hypothesis is correct. Here is one possible hypothesis for this experiment: "All the brine shrimp will be dead by the time the pH level of the water reaches 4.5."

In this case, the **variable** you will change is the pH level of the water, and the variable you will measure is the number of brine shrimp that remain alive. You expect them all to die by the time the pH level reaches 4.5.

You will also set up a **control experiment**. It will be identical to the "real" experiment, except that the pH level will remain the same in the control water and decrease in the experimental water.

What Are the Variables?

Variables are anything that might affect the results of an experiment. Here are the main variables in this experiment:

- the size and health of the brine shrimp
- the number of brine shrimp in a given amount of water
- the temperature of the water
- the kind and amount of food the brine shrimp receive
- the pH level of the water

In other words, the variables in this experiment are everything that might affect the survival of the brine shrimp. If you change more than one variable, you will not be able to tell which variable had the most effect on the shrimps' survival.

After each pH decrease in the experimental water, you will estimate the number of brine shrimp that remain alive in the experimental and the control water. If the shrimp in the experimental water are all dead by the time the pH reaches 4.5, while most remain alive in the control water, you will know your hypothesis is correct.

Level of Difficulty

Moderate, because of the time involved.

Materials Needed

- 1 tablespoon of live brine shrimp (Brine shrimp are sold as fish food at tropical and saltwater fish shops. The clerk will measure 1 tablespoon of shrimp, which contains several hundred shrimp, and pour it into a container of water.)
- 2 wide-mouth jars
- distilled water at room temperature (or tap water that has been in an open container overnight to allow the chlorine in it to evaporate)
- 2 small, clear containers
- 2 labels and a marker
- litmus paper and a color scale
- white vinegar
- measuring spoons

- a stirrer
- 2 medicine droppers
- 1 package dry yeast
- Optional: small aquarium pump with two outlets and plastic tubing

Approximate Budget

$5 for the brine shrimp, litmus paper, and yeast. (The other materials should be available in most households.)

Timetable

One week.

Step-by-Step Instructions

1. Fill both glass jars half-full of water.

2. Use the two small, clear containers to divide the brine shrimp into two equal portions.

3. Pour one portion of shrimp into each of the jars. Rinse the small containers. Label one jar *Control* and one *Experiment*.

Step 4: Recording chart for Experiment 1.

Record of Live Brine Shrimp					
	Mon.	Tues.	Wed.	Thurs.	Fri.
Experiment Jar	pH:	pH:	pH:	pH:	pH:
	No. of Shrimp:	No. of Shrimp:	No. of Shrimp:	No. of Shrimp:	No. of Shrimp:
Control Jar	pH:	pH:	pH:	pH:	pH:
	No. of Shrimp:	No. of Shrimp:	No. of Shrimp:	No. of Shrimp:	No. of Shrimp:

How to Experiment Safely

Be careful in handling the glass jars. If possible, wear goggles so the vinegar will not splash in your eyes.

4. Dip a different strip of litmus paper into each jar, check the color scale, and record the beginning pH level of each jar on a chart like the one illustrated.

5. Use the following steps to take a sample of water from each jar and estimate the number of live shrimp in it:
    a. Gently stir the water in the experimental jar until the shrimp are distributed evenly.
    b. Quickly use a medicine dropper to take out a sample of water and shrimp.
    c. Deposit the sample in one of the clear containers.
    d. Count or estimate the number of live brine shrimp in it.
    e. Record the number on your chart.
    f. Pour the sample back into the same jar.
    g. Rinse the dropper and container.
    h. Complete the same process with the control jar.

6. Use the other medicine dropper to slowly add 2 tablespoons (30 ml) of vinegar to the experimental jar. Again measure and record the pH level in that jar. Do not add vinegar to the control jar.

7. Place both jars in a warm, lighted place where they will not receive direct sun. Add a pinch of dry yeast to both jars as food for the brine shrimp.

Step 5d: Brine shrimp in a small, clear container.

8. Optional: Attach a length of plastic tubing to each outlet on the aquarium pump. Insert one of the tubes into each jar so it rests on the bottom of the jar. Start the pump, which will keep the water gently moving and increase its oxygen content.

9. Each day for a week:
   a. Add another pinch of dry yeast to both jars.
   b. Add 2 more tablespoons of vinegar to the experimental jar.
   c. Measure and record the pH levels of both jars.
   d. Repeat Step 5 to monitor how many live brine shrimp remain in both jars. If no live brine shrimp remain in the experimental jar before the end of the week, end the experiment.

Troubleshooter's Guide

Below are some problems that may arise during this experiment, some possible causes, and ways to remedy the problems.

Problem: All or nearly all the brine shrimp died in both jars.

Possible causes:

1. The shrimp were "old." The fish shop might have kept those shrimp for some time without feeding them. Try again with a fresh batch of shrimp.

2. The water had too much chlorine or other chemicals in it. Try again with water from a different source or let the water sit longer before using it.

3. The yeast polluted the water. Try again, feeding the shrimp much less yeast or not at all.

4. The water became too cold or too hot. Make the necessary adjustments and try again.

Problem: Very few of the shrimp died in the experimental jar.

Possible cause: The pH did not reach a toxic level. Continue the experiment, further decreasing the pH level of the experimental water.

Summary of Results

Use the data on your chart to create a line or bar graph of your findings. Then study your chart and graph and decide whether your hypothesis was correct. At what pH level did the brine shrimp in the experimental jar start to die in greater numbers? At what level were they all dead? Did most of the shrimp in your control jar survive until the end of the week? Write a paragraph summarizing your findings and explaining whether they supported your hypothesis.

Change the Variables

To vary this experiment, consider these possibilities:

- Try hatching your own brine shrimp from eggs bought at a pet shop. The hatched shrimp will be very small, but cheap, available, and plentiful. Or use a plankton net to collect small aquatic organisms from pond water. You may need to use a microscope to monitor them during the experiment.
- Change the water temperature. Put two jars of water with a pH of 4.8 (mildly acid rain) under different temperature conditions to see if the shrimp tolerate acid rain better at higher or lower temperatures.
- Change the type of acid by using lemon juice. It is more acidic than vinegar and will cause the pH level to drop more quickly.

Experiment 2
Acid Rain and Plants: How does acid rain affect plant growth?

Purpose/Hypothesis

In this experiment, you will use cuttings of plants that are easy to grow, such as ivy, philodendron, begonia, or coleus. You will place two cuttings in water with a pH level of 7.0, which is neutral, and two cuttings in water with a pH of 4.0, which is in the range of acid rain. Your goal is to determine how the acidity affects the growth of roots.

Before you begin, make an educated guess or hypothesis about the outcome of this experiment based on your understanding of acid rain. This educated guess, or prediction, is your **hypothesis.** A hypothesis should explain these things:

What Are the Variables?

Variables are anything that might affect the results of an experiment. Here are the main variables in this experiment:

- the type, size, and health of the plant cuttings
- the air temperature where the jars of cuttings are placed
- the amount of sun the cuttings receive
- the pH level of the water

    In other words, the variables in this experiment are everything that might affect the growth of roots. If you change more than one variable, you will not be able to tell which one had the most effect on root growth.

- the topic of the experiment
- the variable you will change
- the variable you will measure
- what you expect to happen

A hypothesis should be brief, specific, and measurable. It must be something you can test through observation. Your experiment will prove or disprove whether your hypothesis is correct. Here is one possible hypothesis for this experiment: "Cuttings placed in water with a pH level of 4.0 will not grow any roots, while cuttings in water with a pH of 7.0 will begin to grow roots during the experiment."

In this case, the **variable** you will change is the pH level of the water, and the variable you will measure is the amount of roots that grow. You expect no roots to grow in the water with a pH level of 4.0.

The cuttings in the water with a pH of 7.0 serve as a **control experiment,** allowing you to observe root growth when the pH of the water remains neutral. After the two-week period of the experiment, if the cuttings in the neutral water have grown roots, but those in the acid water have not, you will know your hypothesis is correct.

Level of Difficulty

Moderate, because of the time involved.

Materials Needed
- 4 small, clear jars
- 4 labels and a marker
- 2 large water containers
- water
- litmus paper and a color scale
- white vinegar
- baking soda
- measuring cups and spoons
- a stirrer
- 2 cuttings each of two easily grown plants, such as ivy, philodendron, begonia, or coleus (Make sure each cutting has the same number of leaves and same amount of stem.)

Approximate Budget
$5 for the plants and litmus paper. (Ask friends, neighbors, or family members for cuttings so you will not need to buy plants, and the other materials should be available in most households.)

Timetable
Two weeks to observe plant growth.

Step-by-Step Instructions
1. Label the four small jars in this way:
   (name of plant 1), neutral
   (name of plant 1), acid
   (name of plant 2), neutral
   (name of plant 2), acid

2. Pour 2 cups of water into each of the large containers.

3. Use the litmus paper and a litmus color scale to measure the pH level of the neutral or control container. It should be 7.0. If it is higher, add a drop or two of vinegar, stir, and check it again. If it

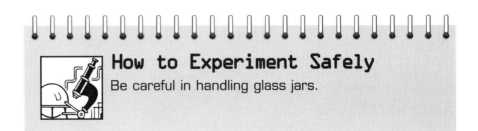

How to Experiment Safely
Be careful in handling glass jars.

Step 6: Plant cuttings in labelled jars of water.

is lower than 7.0, sprinkle in a little baking soda, stir, and check again. Repeat until the color scale shows that the pH level is 7.0.

4. Pour 1 tablespoon (15 ml) of vinegar into the acid or experimental container, stir, and check the pH level. It should be 4.0. If it is higher or lower, add vinegar or baking soda, as in Step 3.

5. Nearly fill the two small jars labeled *Neutral* with the neutral water. Then pour the same amount of acid water into the two small jars labeled *Acid.* Label and save any leftover water so you can keep the small jars full of water with the correct pH level.

6. Place the four plant cuttings in their labeled jars. Make sure the stem and part of the lowest leaf is under water.

7. Place all four jars in a warm, sunny place.

8. Create a chart like the one illustrated. Draw each cutting to show how it looked at the beginning.

9. For the next two weeks:
   a. Every day, make sure all cuttings are still in the water. Add more acid or neutral water to replace any that evaporates. (Be careful to add the right kind to each cup.)
   b. Every other day, check the pH of the water in each cup, and use vinegar or baking soda to adjust it so it is 7.0 or 4.0.

experiment
CENTRAL

Record of Plant Growth

Week 1	Mon.	Tues.	Wed.	Thurs.	Fri.
Acid, Plant 1 Drawing:					
Acid, Plant 2 Drawing:					
Neutral, Plant 1 Drawing:					
Neutral, Plant 2 Drawing:					

Week 2	Mon.	Tues.	Wed.	Thurs.	Fri.
Acid, Plant 1 Drawing:					
Acid, Plant 2 Drawing:					
Neutral, Plant 1 Drawing:					
Neutral, Plant 2 Drawing:					

Troubleshooter's Guide

Below are some problems that may arise during this experiment, some possible causes, and ways to remedy the problems.

Problem: None of the cuttings grew.

Possible causes:

1. The cuttings were infected with insects, fungus, or something else. Try the experiment again with fresh cuttings from different plants. Use different jars or wash the old jars well.

2. The cuttings were from old, woody sections of the plant. Try cuttings from the growing tips of the plants.

3. The cuttings did not receive enough sun or became too cold or too hot. Perhaps their stems did not remain in the water. Try again, placing the cups in a warm (not hot) place where they will receive several hours of sun every day. Check to make sure the stems remain underwater.

Problem: All of the cuttings grew about the same amount.

Possible causes:

1. The pH of the water in the acid jars might not have remained at 4.0. Try the experiment again, carefully checking the pH levels during the observation period.

2. Perhaps both kinds of plants are tolerant of acid water. That would mean your hypothesis is incorrect for these kinds of plants.

c. Every day, record any changes or growth on the chart. Clearly show any roots that grow longer or branch out, leaves that grow larger, and the emergence of new leaves.

Summary of Results

Study the drawings on your chart and decide whether your hypothesis was correct. Did both cuttings in acid water not grow at all? Or did

OPPOSITE PAGE:
Step 8: Recording chart for Experiment 2.

they grow some, but less than those in neutral water? Was the cutting of one plant more tolerant of acid water than the cutting of the other plant? Did both cuttings in neutral water grow as you expected? Write a paragraph summarizing your findings and explaining whether they supported your hypothesis.

Change the Variables

Here are some ways you can vary this experiment:

- Use different kinds of plants.
- Water potted plants with acid and neutral water and compare their leaf and stem growth and appearance, general health, and frequency of blooming, if applicable, over time.
- Use water with different pH levels, such as 5.0, 4.0, and 3.0 to determine if growth decreases with each increase in acidity.

 # Design Your Own Experiment

How to Select a Topic Relating to this Concept

You can explore many other aspects of acid rain. Consider what puzzles you about this topic. For example, what would happen if you added vinegar or another acid to a jar of water with limestone (calcium carbonate) gravel in the bottom? Lime is a base that can neutralize acid, so would the pH level of the water still drop with the limestone in there?

How does ground lime affect plants that have been damaged by acid rain? Will they begin growing well again if lime neutralizes the soil? What if lime is applied first and then the plants are watered with acid rain? Will the lime protect them? How does acid rain affect the germination of seeds? Which plants are more tolerant of acid rain than others?

Check the Further Reading section and talk with your science teacher or school or community media specialist to start gathering information on acid rain questions that interest you.

Steps in the Scientific Method

To do an original experiment, you need to plan carefully and think things through. Otherwise, you might not be sure what question you are answering, what you are or should be measuring, or what your findings prove or disprove.

Here are the steps in designing an experiment:

- State the purpose of—and the underlying question behind—the experiment you propose to do.
- Recognize the variables involved, and select one that will help you answer the question at hand.
- State a testable hypothesis, an educated guess about the answer to your question.
- Decide how to change the variable you selected.
- Decide how to measure your results.

Recording Data and Summarizing the Results

In the two acid rain experiments, your raw data might include not only charts of brine shrimp survival rates and root growth, but also drawings or photographs of these changes.

If you display your experiment, limit the amount of information you offer, so viewers will not be overwhelmed by detail. Make clear your beginning question, the variable you changed, the variable you measured, the results, and your conclusions. Viewers—and judges at science fairs—will want to see how your experiment was set up. You might include photographs or drawings of the steps of the experiment. Viewers will want to know what materials you used, how long each step took, and other basic information.

Related Projects

You can undertake a variety of projects related to acid rain. For example, you might explore how acid rain affects buildings, statues, and other outdoor structures. Which kinds of stone are most susceptible to damage from acid rain? How do people fare in regions with highly acidic rain? Do they have more respiratory problems?

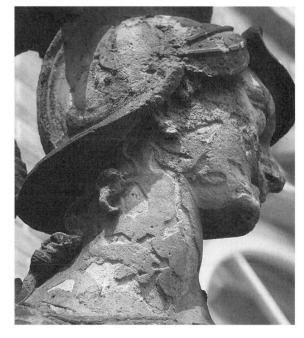

The sulphur in acid rain reacts with the limestone in statues, forming a powder that easily washes away.

For More Information

Asimov, Isaac. *What Causes Acid Rain?* Milwaukee: Gareth Stevens Children's Books, 1992. ❖ Explores the causes, harmful effects, and ways to prevent acid rain.

Edmonds, Alex. *A Closer Look at Acid Rain.* Brookfield, CT: Copper Beech Books, 1997. ❖ Examines the causes of acid rain; its effects on plants, lakes, and human health; and ways to tackle the problem.

Gutnik, Martin. *Experiments That Explore Acid Rain.* Brookfield, CT: Millbrook Press, 1992. ❖ Outlines projects and experiments dealing with acid rain.

Patten, J.M. *Acid Rain.* Vero Beach, FL: Rourke, 1995. ❖ Focuses on the environmental effects of acid rain.

Rainis, Kenneth. *Environmental Science Projects for Young Scientists.* New York: Franklin Watts, 1994. ❖ Outlines detailed projects easily completed by middle school students.

Woodburn, Judith. *The Acid Rain Hazard.* Milwaukee: Gareth Stevens Children's Books, 1992. ❖ Discusses the causes of and damages resulting from acid rain, along with ways to reverse its effects.

chapter 2

Annual Growth

Did you ever measure your height to see how much taller you were than the year before? This change is your annual growth. In humans, annual growth depends on factors such as your age (babies grow at a faster rate than teenagers) and your **genes** (which make sure your growth pattern is similar to that of your parents and grandparents). How can we determine the annual growth of other organisms, and what factors can we find that affect their growth?

Trees are probably the tallest living organisms you will see in your life. Yet most trees around you grew from seeds no larger than the eraser on a pencil. The process by which these tiny seeds become trees is fascinating and easy to observe, when you know what to look for.

How does a tree grow?

A tree grows in two ways. The tips of its branches and tips of its roots contain cells that reproduce, making the tree taller and its roots deeper. Another layer of dividing cells increases the width of the tree's trunk little by little, increasing its support and providing a route for water to reach the upper branches. While a tree is alive, scientists can determine its growth rate by measuring the change in its diameter and also by observing the patterns of new growth on branches and twigs. When a tree has fallen or been cut down, scientists can learn much about the tree's growth throughout its life and can even learn about changes in climate and soil composition long ago by examining the growth rings inside the main trunk.

Words to Know

Alga/Algae:
Single-celled or multicellular plants or plant-like organisms that contain chlorophyll, thus making their own food by photosynthesis. Algae grow mainly in water.

Autotroph:
An organism that can build all the food and produce all the energy it needs with its own resources.

Chlorophyll:
A green pigment found in plants that absorbs sunlight, providing the energy used in photosynthesis, or the conversion of carbon dioxide and water to complex carbohydrates.

TOP: *When a tree has fallen or been cut down, its annual growth rings become visible.*

BOTTOM: *You can learn about a tree's growth pattern by observing the segments of twigs on the tree.*

ⓦords to Know

Cyanobacteria:
Oxygen-producing aquatic bacteria capable of manufacturing its own food; resembles algae.

Dormancy:
A state of inactivity in an organism.

Fungi:
Kingdom of various single-celled or multicellular organisms, including mushrooms, molds, yeasts, and mildews, that do not contain chlorophyll.

Gene:
A segment of a DNA (deoxyribonucleic acid) molecule contained in the nucleus of a cell that acts as a kind of code for the production of some specific protein. Genes carry instructions for the formation, functioning, and transmission of specific traits from one generation to another.

The growth rings that are visible on a tree stump result from the tree's cycle of growth and **dormancy.** The interior of a tree's trunk contains special tube-like vertical cells called **xylem,** which function as a vital part of the tree's water-transport system. Each year, new xylem is produced near the outer layer of bark. In the spring, when conditions are usually wettest, the tree produces large xylem cells. During the drier months of summer, the tree produces smaller xylem cells. In the

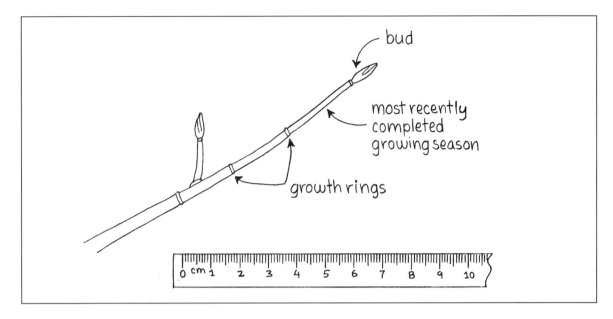

bud

most recently completed growing season

growth rings

0 cm 1 2 3 4 5 6 7 8 9 10

winter, the tree's growth cycle goes into a state of dormancy, a period of inactivity to keep its energy in reserve while water is scarce.

This alternating pattern of fast and slow growth causes the dark and light pattern of rings you can see on the tree stump. Each ring represents a growing season. Generally, a larger, more prominent ring marks a longer, wetter growing season. In this way, scientists have been able to pinpoint when climatic changes occurred long ago in a region's history. A skilled scientist with the right tools can learn even more from a tree's rings, such as when the tree experienced changes in soil composition, forest fires, and floods.

We can also learn about a tree's growth pattern by observing the segments of twigs on the tree. Each spring, the tree will put out a bud at the end of each twig. That bud forms the beginning of that year's new growth. Once the twig grows beyond the point where the bud first formed, the remnants of the bud create a **scar**, or ring. These rings mark off each year of the tree's growth. The most recent segment is the one closest to the end of the twig (assuming the twig has not been broken).

Some twigs exhibit growth rings going back many years. Once a growing season is completed, that season's segment will not grow any longer. The segments can give you a rough indication of how much growth a tree experienced in one season compared to other seasons. Remember that growth may not be the same from one side of a tree to the other, especially in large trees. The segment indicates most accu-

On many trees, twigs exhibit evidence of past growing seasons by the distance separating their scars or annual growth rings.

Words to Know

Heterotrophs:
Organisms that cannot make their own food and that must, therefore, obtain their food from other organisms.

Hypothesis:
An idea in the form of a statement that can be tested by observation and experiment.

Lichen:
An organism composed of a fungus and a photosynthetic organism in a symbiotic relationship.

Niche:
The specific location and place in the food chain that an organism occupies in its environment.

experiment
CENTRAL

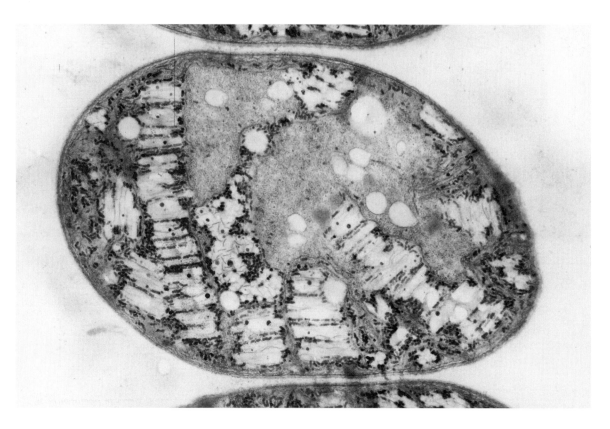

Lichens are actually complex partnerships of different organisms working together. (Peter Arnold Inc. Reproduced by permission.)

rately how much growth occurred on that branch of the tree in a given growing season.

In the first experiment, you will compare the annual growth pattern of twigs on several trees in your area with the rainfall figures for each year. You will then determine if precipitation in your area has had a measurable effect on the trees' annual growth.

Lichens: another kind of annual growth

Have you ever noticed the patches of colorful plant life that sometimes grow on rocks and buildings? Some resemble greenish-brown stains, while others look like blotches of mold. When examined closely, some appear to be tiny forests of hairy branches. These are actually a unique and fascinating life form called **lichens**. Scientists who study lichens are known as lichenologists. One of the most renowned lichenologists was Beatrix Potter, the author of *The Tale of Peter Rabbit*. Though better known for her children's stories, Potter devoted much of her time to the study of lichens and produced detailed watercolor illustrations of different lichen forms.

Words to Know

Photosynthesis:
Chemical process by which plants containing chlorophyll use sunlight to manufacture their own food by converting carbon dioxide and water to carbohydrates, releasing oxygen as a by-product.

Lichens are far more complex than they appear. Each lichen contains two partners, usually a **fungus** and an **alga,** that bond together in a symbiotic relationship. **Symbiosis** occurs when two organisms form a relationship that benefits both. By combining the advantages of fungus with the advantages of algae, the lichen is able to survive where other organisms would perish.

The most visible part of the typical lichen is a fungus. Fungi are plant-like organisms that differ from true plants in that they are **heterotrophs,** organisms that must get their food from other organisms. Fungi usually get their food from dead and decaying matter. Fungi are composed of thin strands that form a network that becomes a home for the fungus's partner. That partner is usually an alga, although some lichens contain **cyanobacteria** instead. If you examine a cross-section of a lichen, you will usually see the alga as a thin layer of green just under the organism's top layer. Algae are tiny plants that use photosynthesis to create nutrients, making them **autotrophs.**

Lichens can survive harsh environments

The symbiotic relationship between the fungus and the algal cells of the lichen depends on the structure and functioning of each. The fungus is capable of securing itself to inhospitable surfaces, such as bare

LEFT: Cyanobacteria are single-celled organisms that perform photosynthesis. (Photo Researchers Inc. Reproduced by permission.)

RIGHT: The fungus in a lichen provides a protecting structure for the algal cells, which provide food.

Words to Know

Symbiosis: A pattern in which two or more organisms live in close connection with each other, often to the benefit of both or all organisms.

Variable: Something that can affect the results of an experiment.

Xylem: Plant tissue consisting of elongated, thick-walled cells that transport water and mineral nutrients.

rock or even plastic. Often, however, a fungus would not find sufficient nutrients in such a habitat. The algal cells, on the other hand, can produce food by photosynthesis, but they could not survive on their own on a bare rock. The two form a symbiotic union. The fungus provides the algae with protection from the harsh environment, while the algae provide the fungus with food.

Cyanobacteria are among the most ancient organisms on Earth. They are usually found in water, sometimes joining together in colonies. Cyanobacteria contain chlorophyll and perform photosynthesis, and thus they sometimes are found in lichens in place of algae.

Lichen growth patterns can be used to determine the age of rocks and rock formations because the rate of growth is extremely slow and regular. Lichens serve as a good indicator of air pollution levels because of their sensitivity to impurities in the atmosphere. In the second experiment, you will utilize two samples of living lichen to measure differences in air quality in different places.

Experiment 1
Tree Growth: What can be learned from the growth patterns of trees?

Purpose/Hypothesis
For this experiment, you will examine and collect growth data from branches of different trees. Then you will determine whether these data correspond to the precipitation in your region. Before you begin, make an educated guess about the outcome of this experiment based on your knowledge of plant growth. This educated guess, or prediction, is your **hypothesis.** A hypothesis should explain these things:
- the topic of the experiment
- the variable you will change
- the variable you will measure
- what you expect to happen

A hypothesis should be brief, specific, and measurable. It must be something you can test through observation. Your experiment will prove or disprove whether your hypothesis is correct. Here is one possible hypothesis for this experiment: "The branches of trees in this area will show similar growth patterns over the past few years because

What Are the Variables?

Variables are anything that might affect the results of an experiment. Here are the main variables in this experiment:

- the species of tree being examined
- the condition of the branches (damaged or not damaged, for example)
- variations in rainfall among the areas where the trees are growing
- the amount of fertilizer or other nutrients each tree receives
- other factors influencing growth, such as the amount of sunlight, level of air pollution, and the presence of disease in the trees

   If you change more than one variable, you will not be able to tell which variable had the most effect on the growth patterns. Try to keep all variables the same except the one you are examining: the amount of annual growth.

they all received the same amounts of rainfall during each growing season."

   In this case, the **variable** you will change is the type of tree, and the variable you will measure is the growth pattern over the past few years. You expect the growth patterns to be similar.

Level of Difficulty
Moderate.

Materials Needed
- sketchbook and pencil
- ruler (one showing millimeters or sixteenths of an inch)
- pruning shears (optional)
- camera (optional)

Approximate Budget
$2.

How to Experiment Safely

If you choose to cut a branch or branches to illustrate your findings, be sure to ask permission before cutting. Use proper protective wear and be careful with the pruning shears.

Timetable

This experiment should be done in two periods of at least 15 minutes each: one period to collect and organize data, and the other to interpret the data and present the results.

Step-by-Step Instructions

1. Choose branches or twigs that exhibit the visible signs of annual growth described above. Select different trees in different locations. If you use fallen branches, make sure they are recently fallen. Otherwise, you will not be sure when the most recent growth occurred. Branches that have been split or damaged, especially at the growth tip, may not provide useful results.

2. Ask your teacher or an adult before cutting any branches or twigs. Remember that any change you make to the natural environment will probably have a lasting effect, so avoid damaging trees whenever possible. If you decide to cut a branch to illustrate your project, do not cut the branch too close to the trunk or greater branch, as this could harm the tree.

3. Note the number of segments you can find on the branch you have selected. Determine which is from the most recently completed growing season and then note how many growth seasons are visible on your branch.

4. Sketch or photograph the branch. Note as much information as possible about the tree and its immediate environment. What kind of tree is it? Is it competing with other trees for water and sunlight? Might there be some other environmental factors affecting its growth, such as air pollution or drainage from parking lots or sidewalks? Check whether the tree is receiving water from an irrigation or sprinkler system. This would have a clear effect on your data, and may make an interesting comparison for your study.

experiment
CENTRAL

Tree Growth Chart

Segment number	Segment year	Length of segment
1	1998	2.3 cm
2	1997	1.9 cm
3		

Step 5: Example of tree growth recording chart.

5. Measure each segment and record your data. Use a chart to keep your information consistent. Your chart should look something like the illustration.

6. Once you have found and examined a number of different samples, use your data to test your hypothesis. For each sample, find the year in which the least growth occurred. Then find the year in which the greatest growth occurred. Use the different samples you have for each growing season to find an average growth for that year. Ask your teacher or librarian for help in finding annual rainfall figures for the years for which you have sample. Compare these figures to the results of your branch measurements.

Summary of Results

Examine your results and determine whether your hypothesis is correct. Did the samples show consistently greater or lesser growth for one or more growing seasons? If so, did those years have more or less rainfall than usual?

Troubleshooter's Guide

Here is a problem that may arise, a possible cause, and a way to remedy the problem.

Problem: The amount of growth varies greatly from tree to tree.

Possible cause: Different types of trees can have drastically different growth rates. Remember that you are looking for which years had the greatest and the least growth for each tree—a factor that may be consistent from tree to tree regardless of each one's growth rate.

Change the Variables

You can vary this experiment by changing the variables. Instead of comparing growth seasons, try simply comparing growth rates from one type of tree to another. See if you can find which tree branches in your area exhibit the most growth in a season. Which tree branches grow the least?

Experiment 2
Lichen Growth: What can be learned from the environment by observing lichens?

Purpose/Hypothesis

For this experiment, you will need to locate different lichens in various habitats around your school and/or home. Counting and measuring the number of lichens you find growing in different areas will give you a rough idea of the amounts of air pollution present. Lichens are nearly everywhere. You will need, however, to find samples large enough to examine and measure. In rural environments, this should not be difficult. Lichens can frequently be found on trees, dead wood, and rocks. Before you begin, make an educated guess about the outcome of this experiment based on your knowledge of lichens. This educated guess, or prediction, is your **hypothesis.** A hypothesis should explain these things:

What Are the Variables?

Variables are anything that might affect the results of an experiment. Here are the main variables in this experiment:

- the species of lichen being examined

- the surface on which the lichen is growing

- the amount of sunlight and rainfall the lichen receives

- the location of the lichen relative to sources of air pollution

In other words, the variables in this experiment are everything that might affect the size and numbers of the lichens. If you change more than one variable, you will not be able to tell which variable had the most effect on lichens.

- the topic of the experiment
- the variable you will change
- the variable you will measure
- what you expect to happen

A hypothesis should be brief, specific, and measurable. It must be something you can test through observation. Your experiment will prove or disprove whether your hypothesis is correct. Here is one possible hypothesis for this experiment: "Fewer and smaller lichens will grow in areas with higher levels of air pollution (near roads and factories) than in areas with cleaner air."

In this case, the **variable** you will change is the location of the lichens, and the variable you will measure is their number and size. You expect fewer and smaller lichens will be found near sources of air pollution.

Level of Difficulty
Moderate.

Materials Needed
- sketchbook and pencil
- magnifying glass

- ruler (one showing millimeters or sixteenths of an inch)
- camera (optional)

Approximate Budget
$5.

Timetable
This experiment requires a commitment of several hours searching and cataloging lichens.

Step-by-Step Instructions
1. Your research for this experiment should begin in the library. It will be worthwhile to photocopy photographs and illustrations of different forms of lichen and bring this information with you when you go out looking for lichen.

2. Remember that lichens can be quite fragile. Treat lichens gently while measuring and sketching them.

3. If you are working together with a group, you might find it useful to divide the responsibilities. Have one group member sketch the lichen while others measure or write brief descriptions of the lichen's habitat. Prepare a chart on which you will record your observations for each lichen you find. Your chart should look something like the illustration.

4. Once you have found a lichen, take note on your chart of the habitat. Is the lichen growing on a tree or rock, or on some other object, such as a rusted barrel? How close is the lichen to the nearest source of air pollution? Note all other environmental factors that might affect the rate of lichen growth, such as shelter from

How to Experiment Safely
This project puts you in contact with fungus from the wild. NEVER eat any wild fungus, even one that looks familiar. Fungi that closely resemble edible mushrooms can in fact be highly toxic. Treat lichens the same way. Though some are edible, many are not. You should also wear gloves when handling the fungus.

Lichen Chart

Location of Lichen	Local Environment	Approx. Height	Approx. Width
① side of building 441 Main St.	close to major road and parking lot	9 cm	7 cm
② tree in Baldwin Park	far from roads (approx. ½ mile)	15 cm	9 cm

Step 3: Example of the lichen recording chart.

rain and sun. Next examine the lichen itself. Describe it as clearly as possible, identifying its color, form, and texture.

5. Measure the lichen using your ruler. Lichen that grow in patches start with one tiny spore-like structure and then grow outward, like mold on bread. Therefore, try to locate the largest single sample instead of measuring two that have grown together. Measure the lichen's greatest horizontal length and greatest vertical length and record this data on your chart.

6. Select different sites that are more likely to show the effects of air pollution. Try to find lichens at different distances from highways, airports, or factories. Roadway intersections often produce increased pollution levels due to cars and trucks stopping and starting.

Summary of Results

Examine your results and determine whether your hypothesis is correct. Did you find a consistent difference in the size (or presence) of lichens on trees closer to roads or parking lots? What other factors did you note that might be affecting lichen growth? Write a summary of your findings.

Troubleshooter's Guide

Here is a problem that may arise, a possible cause, and a way to remedy the problem:

Problem: No lichens can be found.

Possible cause: Some areas, particularly urban environments with high levels of air pollution, may not have any lichens. If you think this may be possible, check with your teacher before attempting this experiment.

Change the Variables

There are several ways you can vary this experiment. Try measuring the effect of changes on the lichen, such as treatment of sunlight and moisture, competition with other plants, or exposure to lichen-eating animals.

 # Design Your Own Experiment

How to Select a Topic Relating to This Concept

Try growing lichen in a controlled environment. If you find lichen growing on an easily movable object, such as a piece of dead wood or a small rock, try carefully moving that rock into your classroom or laboratory. Remember that the lichen needs light and moisture. If you are able to transport lichen, you can design an experiment that will more accurately test the effects of different air qualities on the lichen.

Check the For More Information section and talk with your science teacher or school or community media specialist to start gathering information on annual growth or lichen questions that interest you.

Steps in the Scientific Method

To do an original experiment, you need to plan carefully and think things through. Otherwise, you might not be sure what question you are answering, what you are or should be measuring, or what your findings prove or disprove.

Here are the steps in designing an experiment:

- State the purpose of—and the underlying question behind—the experiment you propose to do.
- Recognize the variables involved, and select one that will help you answer the question at hand.
- State your hypothesis, an educated guess about the answer to your question.
- Decide how to change the variable you have selected.
- Decide how to measure your results.

Recording Data and Summarizing the Results

In the experiments included here and in any experiments you develop, you can try to display your data in more accurate and interesting ways. Collecting samples of the lichen you measure for your experiment will make the results more interesting to viewers. Photographs of the lichen you find can be helpful, but you may discover that careful sketches can reproduce details that are not clear in photographs.

Related Projects

Projects and experiments in annual growth can reveal much about our environment that usually occurs too slowly for us to notice. Some fascinating experiments can be conducted over longer periods of time if you establish a structure for other students to follow later on. Talk with your teacher and classmates about starting a project to monitor long-term tree or lichen growth in your area. Take measurements of the circumference of the tree trunks near your school and record your data for comparison next year. Look for sources of information on tree growth in the past. Old photographs cannot provide exact measurements, but they can show roughly how much a tree has changed over a period of years or even decades.

For More Information

Menninger, Edward. *Fantastic Trees.* Portland, OR: Timber Press, 1995. ❖ A fun and fascinating look at strange and little-known facts about trees.

Platt, Rutherford. *1001 Questions Answered About Trees.* New York: Dover Publishing, 1992. ❖ A question-and-answer format book covering practically everything about trees.

Pollick, Steve. *Find Out Everything About Plants.* London: BBC Publishing, 1996. ❖ Contains a number of interesting and clearly illustrated project ideas on plant- and growth-related topics.

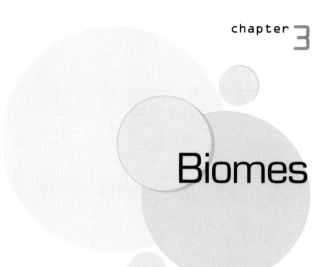

chapter 3

Biomes

If you have ever hiked in a forest or driven through a desert, what you saw was a biome. **Biomes** are large geographical areas with specific climates and soils, as well as distinct plant and animal communities that are all interdependent.

Most biomes are on land. Our oceans make up a single biome. Besides **temperate** forest and desert, the major land biomes include **tundra, taiga** (pronounced TIE-gah), temperate **deciduous** (pronounced deh-SID-you-us) forest, tropical rainforest, and grassland. To understand how biomes work, let us look at some of them.

Into the woods

Maybe you have hiked in a taiga biome, the biome that receives the most snow. Unlike its neighboring biome, the tundra, which is treeless and characterized by low-lying plants, the taiga is sometimes called the **boreal** (pronounced BORE-e-al) **coniferous** (pronounced CONE-if-er-us) forest and is probably the largest of all the land biomes. The taiga biome extends across the northern parts of North America, Asia, and Europe. It is dominated by coniferous, or cone-bearing, trees such as pine, spruce, larch, and fir. These trees resist cold, which is a good thing, because temperatures have been recorded as low as –90°F (–67°C) and reach an average of only 59°F (15°C). The tree roots do not penetrate deeply and tend to interconnect with other tree roots around them. Each tree is basically held down by its neighbors on all sides.

Trees in the taiga biome survive in soil that is frozen for most of the year. Soil moisture comes from melted snow and summer rains,

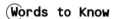

Words to Know

Biomes:
Large geographical areas with specific climates and soils, as well as distinct plant and animal communities that are interdependent.

Boreal:
Northern.

Coniferous:
Refers to trees, such as pines and firs, that bear cones and have needle-like leaves that are not shed all at once.

Pine, spruce, and fir trees form part of the taiga biome. (Photo Researchers Inc. Reproduced by permission.)

Words to Know

Deciduous:
Plants that lose their leaves during some season of the year, and then grow them back during another season.

Desert:
A biome with a hot-to-cool climate and dry weather.

Desertification:
Transformation of arid or semiarid productive land into desert.

Ecosystem:
An ecological community, including plants, animals and microorganisms considered together with their enviroment.

Ephemerals:
Plants that lie dormant in dry soil for years until major rainstorms occur.

Fungus (fungi):
Various single-celled or multicellular organisms, including mushrooms, molds, yeasts, and mildews, that do not contain chlorophyll.

but during the winter, the cold temperatures make water absorption difficult because the ground is frozen. So these trees have built-in adapters to help them survive. For example, spruce and fir trees have long, thin, wax-covered needles. The waxy surface acts as an insulator, helping them retain water and heat. Snow slides off more easily, avoiding branch breakage. These needles conduct photosynthesis so efficiently that they can make food even during winter, when the Sun's rays are weaker.

Trees are not the only inhabitants of this biome. About fifty species of insects, including mites, live here. Moose, snowshoe hares, deer, and elk make their home in the taiga as well as wolves, porcupines, lynxes, and martens, who roam the taiga during the summer. Seeds from the cones of the trees are food for red squirrels and for birds such as crossbills and siskins.

Life in the desert—with air conditioning

If you drive through a **desert,** do you see much life from your car window? Do not be fooled. There is more living here than just cactus plants. Desert biomes are on every major continent and cover more than a fifth of Earth's surface. While these biomes receive less than 10 inches (25 centimeters) of rainfall a year, with temperatures that range from 75°F (23°C) to 91°F (32°C), desert plants and animals thrive here. Deserts can usually be found in the centers of continents and in the rain shadows of mountains.

Low-growing bushes in Monument Valley are part of this biome's vegetation. (Photo Researchers Inc. Reproduced by permission.)

Lizards, snakes, and other animals pop up at sundown when the soil is cool, then wriggle back into their habitats when the temperature becomes too chilly. They can reappear again at dawn, remaining until the temperature gets too hot. Some of the rodents and other animals that burrow under the soil actually enjoy a kind of underground air-conditioning. They form elaborate tunnels where the Sun's heat cannot penetrate. And moisture from the animals' exhaled breath cools the air and makes their burrows a comfortable 85°F (29°C). Kangaroo rats in the American Southwest and the gerbils of North African and Asiatic deserts choose foods that reduce the amount of water needed for digestion. These rodents can actually absorb water from their urine before excreting wastes.

Many desert plants are **xerophytes** (pronounced ZERO-fights), plants that require little water to survive. There are also **ephemerals** (pronounced eh-FEM-er-als), plants that can suspend their life processes for years when the soil becomes too dry. When major rainstorms occur, they burst into life. **Succulents** are another type of plant. They retain water in thick fleshy tissues. Birds use the giant saguaro (pronounced sah-GWA-ro; from the Spanish word for the Pima Native American name of this plant) cactus, a succulent plant that grows 50 feet (15 meters) high, as nesting and resting areas in place of trees.

The saguaro cactus is a good example of the interdependence that takes place in a biome. Red-tailed hawks use the branches to nest.

Words to Know

Hypothesis:
An idea in the form of a statement that can be tested by observation and/or experiment.

Succulent:
Plants that live in dry environments and have water storage tissue.

Taiga:
A large land biome mostly dominated by coniferous trees.

Temperate:
Mild or moderate weather conditions.

Hollowed-out trunk and arm spaces are a home for elf owls and gila woodpeckers. The cactus fruits are eaten by rodents, birds, and bats.

Why save the rainforests?

Many people are concerned about saving rainforests because these biomes contain a large number of unique plants. Several acres of rainforest in Borneo may contain 700 different species of trees. More than 50,000 plant species make their home in the rainforests of the Amazon Basin in South America. Up to eighty different species of plant life might grow on one tree. Tropical rainforests are found only in regions north of the equator on the Tropic of Cancer and south of the equator in the Tropic of Capricorn. Destroying the rainforests reduces the diversity of life on Earth.

If you have ever been in a steamy greenhouse, then you can imagine what a rainforest is like. Warm temperatures average 75°F (23°C) and humidity peaks at a dripping 90 percent for days at a time. This climate encourages an explosion of plant life that supports many different animals. Some scientists estimate that half the living species on Earth live in the rainforests.

Constructing your own mini-biome will help you understand some of the major factors that influence these important areas of life and can cause them to survive or fail.

ⓦords to Know

Tundra:
A treeless, frozen biome with low-lying plants.

Variable:
Something that can affect the results of an experiment.

Xerophytes:
Plants that require little water to survive.

experiment
CENTRAL

Project 1
Building a Temperate Forest Biome

Purpose
Biomes are strongly influenced by the climate and soil type in a particular region. These same factors determine the success of a mini-biome model. In this project, you will attempt to build, grow, and maintain a temperate forest biome. This particular biome is characterized by a temperature range of 32 to 68°F (0 to 20°C). It has an annual precipitation of 20 to 95 inches (50 to 240 centimeters) and a fairly deep soil layer. The purpose of this project is to try to maintain the correct climate, soil, and vegetation in the temperate forest biome.

Level of Difficulty
Moderate. (This project requires continuous tending and attention to maintain a proper climate.)

Materials Needed
- 10-gallon fish tank (plastic, if possible, for safety)
- indoor/outdoor thermometer
- watering container
- gravel
- sand
- topsoil
- incandescent light fixture with a 40-watt bulb (optional)
- plants and/or seeds (choose oak, maple, sassafras, hickory, tulip trees, sweet gum, dogwood)

Note: Choose all deciduous trees. Seeds may be hard to grow unless they have been chilled. If you use trees, they should be very small saplings.

Approximate Budget
$25. (Try to use an old fish tank if possible.)

Timetable
One hour to set up the project and at least six months to maintain the trees and observe changes.

Materials for Project 1.

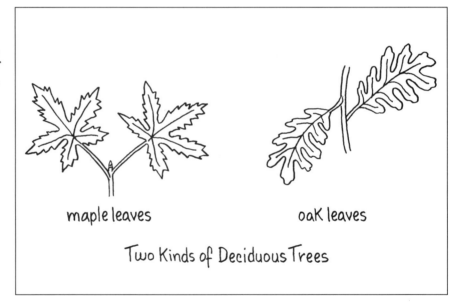

Maple and oak leaves, examples of deciduous trees.

maple leaves   oaK leaves

Two Kinds of Deciduous Trees

How to Work Safely

Ask for assistance when carrying and lifting the fish tank. Do not leave the light fixture on for more than 10 hours at a time.

experiment
CENTRAL

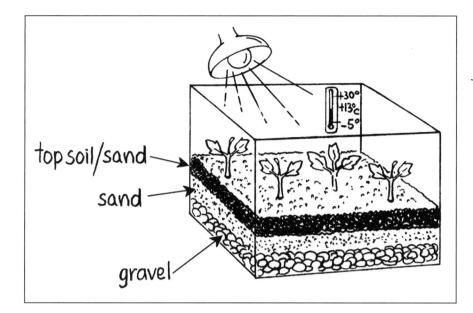

top soil/sand

sand

gravel

Steps 1 to 5: Set-up for fish tank with plants and overhead light.

Step-by-Step Instructions

1. Place a 1-inch (2.5-centimeter) layer of gravel on the bottom of the fish tank.

2. Place a 1-inch (2.5-centimeter) layer of sand over the gravel.

3. Mix 2 parts of topsoil to 1 part of sand. Place a 2- to 3-inch (5- to 7.5-centimeter) layer of the sand/topsoil mixture over the sand layer.

4. Plant 4 to 6 trees. Be sure to cover all the roots. If seeds are being used, place them 1 inch (2.5 centimeters) down in the soil and allow 1 month for them to sprout.

5. Place the thermometer inside the terrarium against the back wall.

6. Water gently until approximately 0.25 inch (0.6 centimeter) of water has accumulated in the gravel layer.

7. Place the fish tank outside or in a sunny place indoors. You must maintain the temperature of the fish tank in the range of 32 to 60°F (0 to 15°C). If you need to provide artificial light, place the incandescent fixture above the fish tank and provide 5 to 10 hours of light per day.

8. Check the project daily, and maintain 0.25 inch (0.6 centimeter) of water in the gravel.

9. Record the growth of the plants and the temperature range.

Troubleshooter's Guide

When you are building a natural environment, many forces of nature can affect the experiment. These include fungus, insects, and too much or too little water. Here are some common problems and a few tips to maintain the best environment.

- Mushrooms, a kind of **fungus,** may grow. Water less, but never allow the soil to dry completely.

- Pests such as insects and spiders may make this biome their home. If they are eating the plants, remove the pests. If not, keep them. They are performing their natural role in the **ecosystem.** Their presence is a sign of a healthy biome.

- Drastic temperature changes overnight can kill the plants. Do your best to maintain an acceptable climate in the fish tank. You may have to move it inside or place it in a shady spot outside, protected from too much rain.

Summary of Results

Graph the data you have collected over the six-month period. The overall growth of the plants will demonstrate the health of the biome environment.

Project 2
Building a Desert Biome

Purpose

In this project, you will build, grow, and maintain a desert biome. The desert biome is characterized mainly by its lack of water, which causes harsh growing conditions. Maintaining the right climate, soil, and vegetation is the goal. This particular biome is characterized by a temperature range of 23 to 60°F (6 to 30°C).

Level of Difficulty

Moderate to difficult because of the length of time needed for the project.

Materials Needed

- 10-gallon fish tank
- indoor/outdoor thermometer
- watering container
- gravel
- sand
- topsoil
- incandescent light with 60-watt bulb
- succulent plants, such as jade plant, strawberry cactus, barrel cactus, etc.

Note: Most plants are easily found in local nursery stores selling houseplants.

Approximate Budget

$25. (Try to get an old fish tank to use.)

Timetable

One hour to set up the project and at least six months to maintain the plants and observe changes.

Step-by-Step Instructions

1. Place a 1-inch (2.5-centimeter) layer of gravel in the bottom of the fish tank.

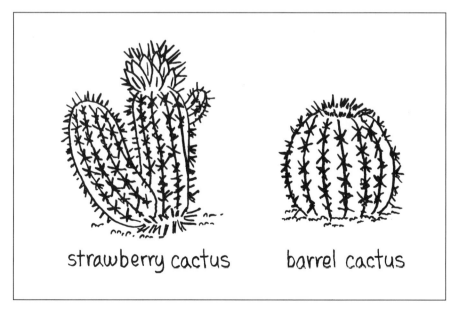

Strawberry and barrel cactuses.

How to Work Safely

Ask for assistance when moving the fish tank. Do not leave the light fixture on for more than 10 hours at a time, as it will get too hot.

2. Mix 1 to 2 cups of topsoil with 6 to 10 cups of sand. Place this mixture over the gravel layer.

3. Place 2 inches (5 centimeters) of sand over the sand/topsoil layer.

4. Plant the cactus and succulents in the fish tank and cover the roots completely.

5. Place the thermometer inside the fish tank, against the back wall.

6. Water sparingly. Pour 2 cups of water on the sand to start. Water the fish tank with 1 cup of water every week after that.

7. Place the light fixture above the fish tank and leave it on for 8 to 10 hours a day.

8. Check the fish tank daily. Record any differences in the plants' growth and in the temperature range.

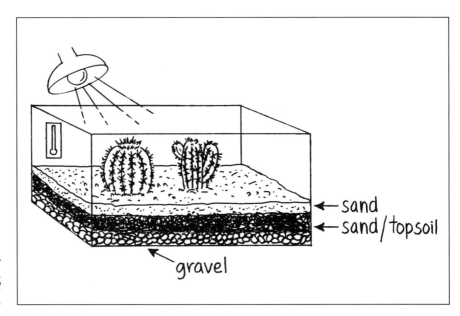

Steps 1 to 5: Set-up for fish tank with cactus, light, thermometer, and sand and gravel layers.

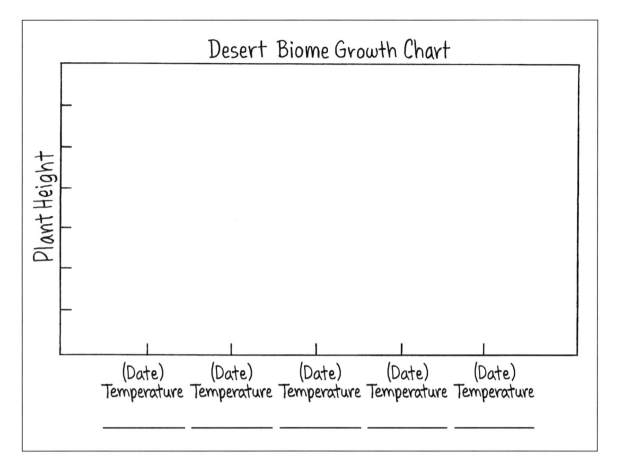

Summary of Results

Graph the data you collected during the project, as illustrated in the Desert Biome Growth Chart. You will notice very little change, as the plants have a very slow growth cycle.

Desert Biome Growth Chart.

Troubleshooter's Guide

In this model, climate conditions are designed to be extreme. The plants have special adaptations to adjust. If insects become a problem, remove them.

Design Your Own Experiment

How to Select a Topic Relating to This Concept

The fish tank projects are models of what takes place in a biome. Many plants and animals have specific adaptations that are suited to that biome or region. What happens when you change the climate of a biome? How does the introduction of a plant from a different biome affect the other plants? There are many experiments you could design to investigate the interactions of plants and animals with their biomes.

Check the For More Information section and talk with your science teacher or school or community media specialist to start gathering information on biome questions that interest you.

Steps in the Scientific Method

To do an original experiment, you need to plan carefully and think things through. Otherwise, you might not be sure what question you are answering, what you are or should be measuring, or what your findings prove or disprove.

Here are the steps in designing an experiment:

- State the purpose of—and the underlying question behind—the experiment you propose to do.
- Recognize the **variables** involved, and select one that will help you answer the question at hand.
- State a testable **hypothesis**, an educated guess about the answer to your question.
- Decide how to change the variable selected.
- Decide how to measure the results.

Recording Data and Summarizing the Results

It is important to document as much information as possible about your experiment. Part of your presentation should be visual, using charts and graphs. Remember, whether or not your experiment is successful, your conclusions and experiences can benefit others.

Related Projects

More specific projects can be performed to get more detailed information about biomes. For instance, scientists are finding that many rainforests are getting drier. Also, a phenomenon called **desertification** has been occurring, turning naturally dry land into desert. Try an experiment in desertification, reducing water to see what happens.

For More Information

Morrison, Marion. *The Amazing Rain Forest and Its People.* New York: Thompson Learning, 1993. ❖ Provides a good summary of this ecological community and how interdependency affects this biome.

Rainis, Kenneth. *Environmental Science Projects for Young Scientists.* New York: Franklin Watts, 1994. ❖ Describes biome and related projects for young people.

Sayre, April Pulley. *Taiga.* New York: Twenty-First Century Books, 1994. ❖ Explores the taiga biome, its animals, plant life, the people who live there, and their impact.

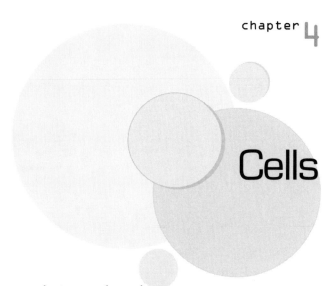

Cells

What do slimy earthworms, majestic lions, and giant redwood trees have in common with humans? We all have cells, tiny units of life that grow and duplicate, gather fuel and building materials, and make energy. **Cells** are present in all living things. Some living things, such as bacteria and some plants, consist of only one cell where all the functions of life take place. They are known as **unicellular.** The average human has 50 to 100 trillion cells. Living things with a great many cells that are joined together are called **multicellular.**

Looks like a monk's cell to me

All humans begin life as a single cell. It weighs no more than a millionth of an ounce. The naked eye cannot see anything that tiny. So no one could have known cells existed until the compound microscope was invented in the late sixteenth century. Between 1590 and 1609, Dutchmen Hans Janssen, his son Zacharias, and Hans Lippershey designed several compound microscopes. In a compound microscope, two or more lenses are arranged to produce a greatly enlarged image.

In 1660, a Dutch drape maker named Anton van Leeuwenhoek (1632–1723) used a microscope to peer at his textiles. He began studying the invisible worlds of nature. Leeuwenhoek designed 250 different microscopes to further his studies. Around that time, Robert Hooke (1635–1703), an English scientist, slid a piece of cork under a microscope. The mass he saw seemed to be made of chambers, like monks' cells in a monastery. He called these chambers "cells."

Words to Know

Cells:
The basic unit for living organisms; cells are structured to perform highly specialized functions.

Cell membrane:
A thin-layered tissue that surrounds a cell.

Cell theory:
All living things have one or more similar cells that carry out the same functions for the living process.

Chloroplasts:
Small structures in plant cells that contain chlorophyll and in which the process of photosynthesis takes place.

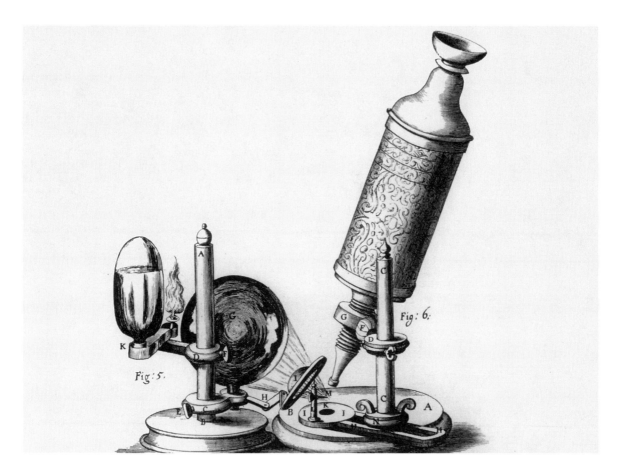

Developing the cell theory

Hooke's cells were from a cork tree's dead and dry bark. The fact that cells are units of life was not understood until the nineteenth century. Between 1838 and 1839 Theodor Schwann and Matthias Schleiden, both German zoologists, independently said that all living things have one or more cells, that all cells are similar, and that in order to exist, these cells carry out the same functions. These facts are now called the **cell theory.** The study of cells is called **cytology.** Rudolf Virchow, a German pathologist, took the cell theory further in 1855 and suggested that all cells are formed by the division of pre-existing cells. Without the cell theory we would never know how organisms grow and develop. We could not treat diseases or pains in our joints, for instance, without knowing what cells do and how they function.

ⓦords to Know

Cytology:
The branch of biology concerned with the study of cells.

What's in there?

Cells are not lifeless blobs. Chemical changes within each cell accomplish many functions, including digestion and breathing. There are

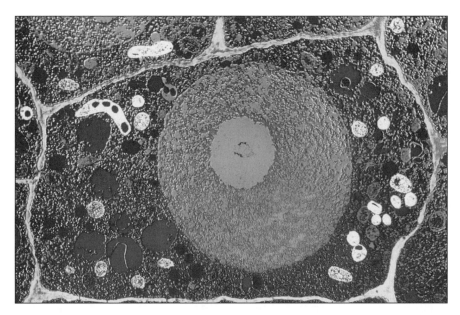

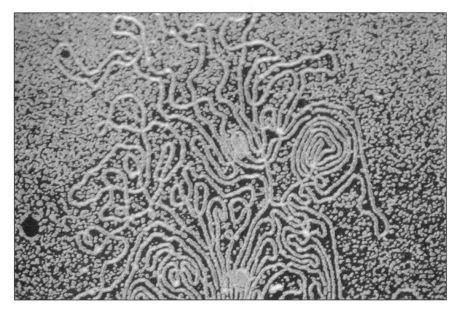

ⓦ Words to Know

Cytoplasm:
The semifluid substance inside a cell that surrounds the nucleus and the other membrane-enclosed organelles.

Dicot:
Plants with a pair of embryonic seeds that appear at germination.

DNA:
Large, complex molecules found in nuclei of cells that carry genetic information for an organism's development.

Embryonic:
The earliest stages of development.

Germination:
The beginning of growth of a seed.

Golgi body:
An organelles that sorts, modifies, and packages molecules.

Hypothesis:
An idea in the form of a statement that can be tested by observation and/or experiment.

two basic types of cells, plant cells and animal cells. Almost all cells share similar features, such as a **cell membrane,** which surrounds the cell. The cell membrane is a thin wall that lets gases, such as oxygen, and fluids, such as nutrients, pass through. **Cytoplasm** (pronounced CY-tow-pla-sim) is the gray, jellylike substance inside the cell membrane. It consists mostly of water but also has many other substances important for cells to function.

Think of a cell as a factory with each division performing specific jobs. **Organelles** (pronounced OR-gan-ells) in the cytoplasm represent those divisions. For instance, **Golgi bodies** are organelles that act as the cleaning crew. Golgi absorb waste, package it up, and send it out for disposal. **Vacuoles** (pronounced VAC-u-ols) are organelles that act as the storage crew. They store food, waste, and chemicals. While there are similarities in cells, there are differences between plant and animal cells. The cytoplasm of plants, for example, contains **chloroplasts,** which gives the plant the ability to make its own food.

It's what makes your hair curly

The nucleus, another organelle, is the cell's library. It lies in the center of the cell and contains DNA. **DNA,** an abbreviation for deoxyribonucleic acid, are molecules that store information. They tell each cell how to develop into a nerve cell, a blood cell, and so on. What makes you unique, as well as what makes you similar to other people, was programmed into your DNA. Each cell contains many strands of DNA. If you put them all together, they would stretch thousands of miles.

Cells are like little companies. They contain tiny workers with functions that help the living organism survive. A company's main goal is to make a profit. A cell's main goal is sustaining life. Conducting projects with a microscope will enable you to see the way in which cells function as a life force.

Project 1
Investigating Cells: What are the differences between a multicellular organism and a unicellular organism?

Purpose

In this project, you will collect, prepare, mount, and compare cells from a multicellular organism and a single-celled **protozoan.** This will allow you to observe the differences between these two basic forms of organisms.

Level of Difficulty

Moderate/difficult, because it requires the use of a compound microscope. (If you are unfamiliar with its use, please ask a teacher or other adult for assistance.)

Monocot: Plants with a single embryonic leaf at germination.

Multicellular: Living things with many cells joined together.

Organelles: Membrane-bounded cellular "organs" performing a specific set of functions within a cell.

Pnematocysts: Stinging cells.

Protozoan: Minute, one-celled animals.

Unicellular: Living things that have one cell. Protozoans are unicellular, for example.

Vacuoles: A part of plant cells where food, waste, and chemicals are stored.

Variable: Something that can affect the results of an experiment.

How to Work Safely

Use caution when collecting cells with toothpicks. When carrying the compound microscope, use two hands. After collecting pond water, wash your hands. Be careful not to stain your clothes or furniture when using the iodine.

Materials Needed

- compound microscope (try to borrow one from a school or friend)
- slides and cover slips, glass or plastic (Note: If your slides are plastic, use plastic cover slips.)
- stain (iodine from drugstore is good; avoid any solution with alcohol, as it will kill any organisms)
- toothpicks (flat-end toothpicks work best)
- eye dropper
- small jar filled with pond water, the dirtier the better

Approximate Budget

$10 for stain, slides, cover slips, and eye dropper.

Timetable

About 1 hour.

Step-by-Step Instructions

1. Use the flat end of a toothpick to gently scrape the inside of your cheek. Don't press too hard! Scrape gently five to ten times.

2. Smear the cells from the end of the toothpick onto a clean slide.

3. Place one drop of stain onto the slide, covering the cells.

4. Gently place the cover slip over the cell culture. (Hint: Gently rest one side on the slide and slowly lower the cover slip until it rests flat.)

5. Examine the slide under the microscope, using low power.

6. Draw what you see and label any parts you recognize.

7. Place two drops of pond water on the center of the slide.

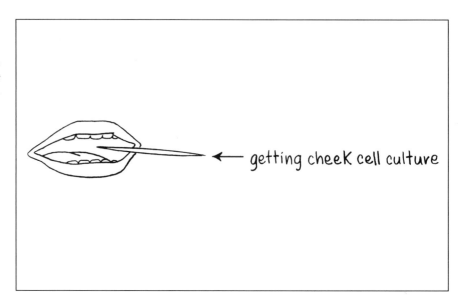

getting cheek cell culture

LEFT: Step 4: Cell culture slide, with cover slip tipping over the cell culture.

RIGHT: Step 7: Pond water cheek cells on low power.

8. Place a drop of stain on the pond water drops.

9. Place the cover slip over the slide using the same technique as with the cheek cells.

10. Examine the slide under the microscope, using low power.

11. Draw what you see and label the parts.

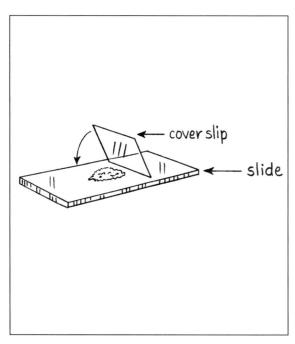

cover slip

slide

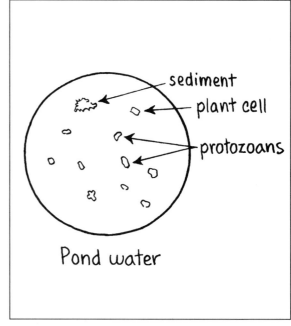

sediment

plant cell

protozoans

Pond water

Troubleshooter's Guide

Problem: Nothing appears on the slide.

Possible cause: You are probably out of focus. Place a small piece of paper on the slide and focus until it is clear. Use the fine focus knob.

Summary of Results

Compare your diagrams and data of the cheek cells and protozoans from the pond water. Determine which cells had a more complex structure. Record a list of the differences between cheek cells and protozoan cells. Note differences such as movement, shape, presence of a cell membrane, and the presence of other cell stuctures. Summarize your observations with sketches and in writing.

Project 2
Plant Cells: What are the cell differences between monocot and dicot plants?

Purpose

In this experiment, you will collect, prepare, and mount cells from two multicellular plants. The multicellular plants you will be working with are **monocot**, that is, plants with a single **embryonic** leaf at **germination**, and **dicot**, plants with a pair of embryonic leaves at germination.

Level of Difficulty

Moderate/difficult, because it requires the use of a compound microscope. (If you are unfamiliar with its use, please ask a teacher or other adult for assistance.)

Materials Needed

- compound microscope (try to borrow one from a school or friend)
- slides and cover slips, glass or plastic (Note: If you are using plastic slides, use plastic cover slips.)
- single-edge razor blade
- thread spool

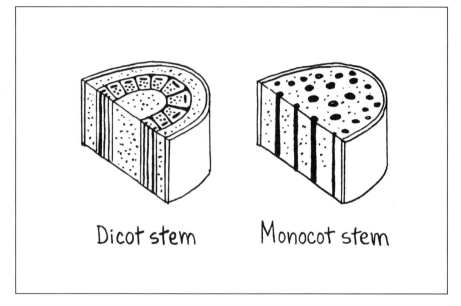

Basic differences between monocot and dicot stems: Dicot stem cells are more orderly; monocot stem cells are more random.

Dicot stem   Monocot stem

- plant stems—tulip and daisy preferred (Go to a local florist and ask for a clipping of the stem.)

Approximate budget
$6 for the slides and cover slips.

Timetable
About 1 hour.

Step-by-Step Instructions

1. Push the tulip stem through the hole in the thread spool until it pokes out the opposite end.

2. Have an adult use the razor to trim the tulip stem, flush to the thread spool. Discard the trimmed piece.

3. Push the tulip stem through the thread spool so about .03 inch (1.0 millimeter) of stem is exposed.

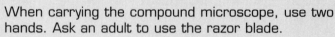

How to Work Safely

When carrying the compound microscope, use two hands. Ask an adult to use the razor blade.

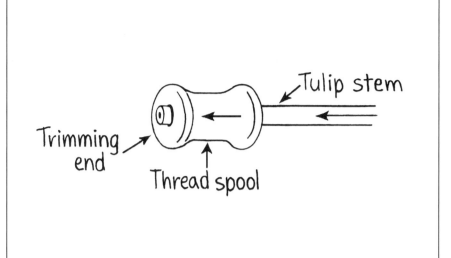

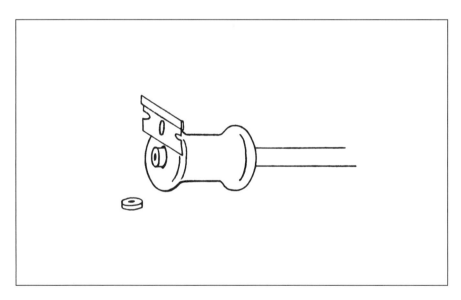

Step 4: Slice of tulip stem trimmed off spool by razor.

4. Carefully using the razor, trim the .03 inch (1 millimeter) of tulip stem flush to the thread spool. Save the trimmed stem.

5. Place the stem slice on the slide and cover with the cover slip.

6. Place the slide on the microscope and examine under low power. Record your observations using drawings and descriptions.

7. Repeat steps 1 to 6 for the daisy stem.

8. Record and compare your observations.

Troubleshooter's Guide

Problem: You cannot see anything.

Possible cause: The stem is too thick. Try cutting the plant stem thinner so the light passes through it.

Summary of Results

Compare your diagrams of the tulip and daisy stems. Which stem had cell patterns that were more orderly? Which stem had more random patterns? A tulip is a monocot, and a daisy is a dicot. Can you tell the difference between monocot and dicot plants by examining their stems?

 Design Your Own Experiment

How to Select a Topic Relating to This Concept

If you choose a topic in biology, you can generally involve the topic of cells. For example, you may be interested in jellyfish and sea anemones. These two creatures share a type of stinging cell called a **pnematocyst,** which paralyzes and kills their prey. The small differences in cell structure give rise to different behaviors and structure of animals and plants.

Check the For More Information section and talk with your science teacher or school or community media specialist to start gathering information on cell questions that interest you.

Steps in the Scientific Method

To do an original experiment, you need to plan carefully and think things through. Otherwise, you might not be sure what question you are answering, what you are or should be measuring, or what your findings prove or disprove.

Here are the steps in designing an experiment:

- State the purpose of—and the underlying question behind—the experiment you propose to do.
- Recognize the **variables** involved, and select one that will help you answer the question at hand.

- State a testable **hypothesis,** an educated guess about the answer to your question.
- Decide how to change the variable you selected.
- Decide how to measure your results.

Recording Data and Summarizing the Results

Your experiment can be useful to others studying the same topic. When designing your experiment, develop a simple method to record your data. This method should be simple and clear enough so that others who want to do the experiment can follow it. Your final results should be summarized and put into simple graphs, tables, and charts to display the outcome of your experiment.

Related Projects

Creating a project about cells offers endless possibilities. You can create a slide collection of cells from many different plants (stem, seed, leaf, needle, root, etc.). You can create a model of a cell labeling the parts and functions. Making a model from colored plastic clay is inexpensive and informative.

For More Information

Bender, Lionel, *Atoms and Cells.* Glouster, ME: Glouster Press, 1990 ❖ Provides background and functions of atoms and cells.

Young, John K. *Cells: Amazing Forms and Functions.* New York: Franklin Watts, 1990. ❖ Good, understandable overview of these units of life for young people.

chapter 5

Chemical Energy

Chemical energy is the energy stored within the bonds of **atoms**. A **bond** is the force that holds two atoms together. Different substances have bonds held together by different amounts of energy. When those bonds are released, a **chemical reaction** takes place, and a new substance is created. Chemical reactions that break these bonds and form new ones sometimes release the excess energy as **heat** and sometimes absorb heat energy from whatever is around them.

Thus, heat energy can be produced or absorbed during a chemical reaction. Reactions that release heat energy are called **exothermic.** Reactions that take in heat energy from the surrounding environment are called **endothermic.** Whether heat energy is given off or absorbed during a chemical reaction depends on the bonds that hold the atoms together.

In a chemical reaction, the original substances are called **reactants.** The new substances that are formed are called **products.** When the bonding structure of the products requires less energy than the bonding structure of the reactants, the excess energy may be released as heat. When the bonding structure of the products requires more energy than the structure of the reactants, it gets that energy by removing heat from its surroundings.

For example, when iron rusts, the iron atoms are combining with oxygen **molecules** in the air to form iron oxide. The chemical reaction of rusting breaks the bonds in the oxygen molecules, releasing heat energy. The bonds between the oxygen atoms and the iron atoms do

Words to Know

Atom:
The smallest unit of an element, made up of protons and neutrons in a central nucleus surrounded by moving electrons.

Bond:
The force that holds two atoms together.

Chemical energy:
Energy stored in chemical bonds.

Chemical reaction:
Any chemical change in which at least one new substance is formed.

The chemical reaction that occurs when iron rusts actually gives off small amounts of heat energy. (Photo Researchers Inc. Reproduced by permission.)

not require as much heat energy as the bonds within the oxygen molecules, so a small amount of energy is released, making the reaction exothermic. The amount of heat released is quite small, and the reaction is normally quite slow, so rusting iron does not feel hot to us. Yet, the energy released can be measured with a thermometer. In the first experiment, you will observe the change in temperature resulting from rusting.

Some exothermic reactions are quite common. One is **combustion,** the burning of organic substances during which oxygen is used to form carbon dioxide and water vapor. The substances formed (ashes, for example) hold less heat energy than the substances burned held. The excess energy is released as heat. The reactions between some chemicals, such as aluminum oxide and iron oxide, can produce great amounts of heat. This reaction is used to produce very high temperatures for industrial purposes.

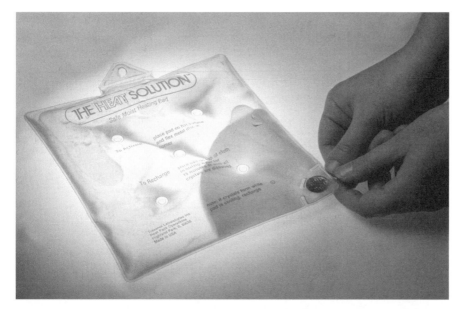

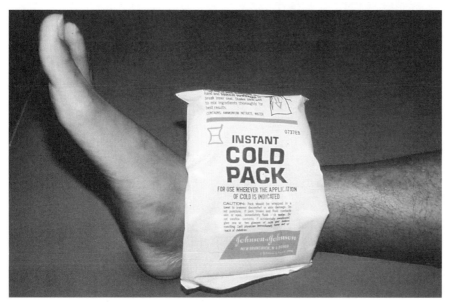

Words to Know

Molecule:
The smallest particle of a substance that retains all the properties of the substance and is composed of one or more atoms.

Product:
A compound that is formed as a result of a chemical reaction.

Reactant:
A compound present at the beginning of a chemical reaction.

Variable:
Something that can affect the results of an experiment.

Endothermic reactions are more rare in nature, but scientists have found ways to create them. For example, an endothermic reaction occurs when you use a chemical cold pack. These packs contain a chemical in powder form that reacts with water. Squeezing the pack breaks down the wall separating the powder from the water. The reaction that occurs absorbs more energy than it releases, making the pack feel cold to you. In the second experiment, you will compare four chemical reactions and determine whether each one is exothermic or endothermic.

Experiment 1
Rusting: Is the chemical reaction exothermic, endothermic, or neither?

Purpose/Hypothesis

In this experiment, you will measure the heat energy released or absorbed by the chemical reaction of rusting, the transformation of iron and atmospheric oxygen into iron oxide. Before you begin, make an educated guess about the outcome of this experiment based on your knowledge of rusting. This educated guess, or prediction, is your **hypothesis.** A hypothesis should explain these things:

- the topic of the experiment
- the variable you will change
- the variable you will measure
- what you expect to happen

A hypothesis should be brief, specific, and measurable. It must be something you can test through observation. Your experiment will

What Are the Variables?

Variables are anything that might affect the results of an experiment. Here are the main variables in this experiment:

• the type of reactants used (iron in the pads and atmospheric oxygen)

• the temperature of the environment in which the samples are tested

• the number of rusting pads in each cup

In other words, the variables in this experiment are everything that might affect the temperature in the cup. If you change more than one variable, you will not be able to tell which variable had the most effect on the temperature.

prove or disprove whether your hypothesis is correct. Here is one possible hypothesis for this experiment: "A rise in air temperature will show that rusting is an exothermic reaction."

In this case, the **variable** you will change is the number of rusting pads in each cup, and the variable you will measure is any change in air temperature. You expect the temperature to go up in the cups with rusting pads and the temperature to go up the most in the cup with the most pads.

As a **control experiment,** you will leave one cup empty and monitor any change in temperature there. If the temperature is higher in the cup with more pads and does not change in the empty cup, your hypothesis will be supported.

Level of Difficulty
Moderate.

Materials Needed
- 4 large Styrofoam cups
- aluminum foil
- 7 steel wool pads (not pads treated with detergent or soap)
- vinegar
- 4 digital laboratory thermometers
- rubber or surgical gloves
- paper towels
- large bowl

Approximate Budget
$10. (If four thermometers are unavailable, the four parts of this experiment can be performed separately with one thermometer.)

Timetable
About 45 minutes.

Step-by-Step Instructions
1. Line the inside of each of the four cups with aluminum foil.

2. Place the seven steel wool pads in the large bowl and soak them thoroughly in vinegar (to remove any coating and encourage rusting). Blot them dry with paper towels.

3. Place one pad in the first cup, two pads in the second cup, and four pads in the third cup. The fourth cup will be empty—your control.

How to Experiment Safely

Wear protective gloves when handling the steel wool and vinegar.

4. Push the bulb of one thermometer gently into the steel wool in the first cup. Do not push the bulb down to or near the bottom of the cup. Cover the opening of the cup with aluminum foil. The stem on the thermometer must be visible.

5. Repeat Step 4 for the second, third, and fourth (control) cups.

6. Place all four cups where no other heat sources will affect their temperature.

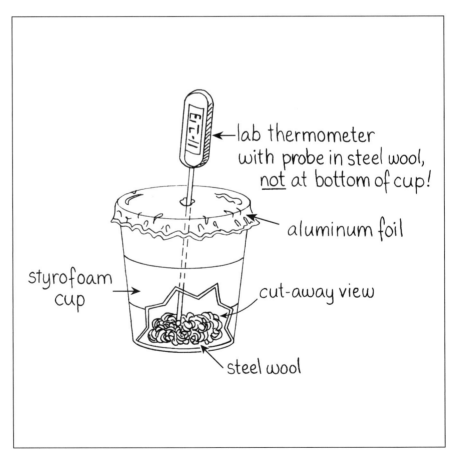

←lab thermometer with probe in steel wool, not at bottom of cup!

aluminum foil

styrofoam cup

cut-away view

steel wool

Step 4: Illustration of rusting set-up.

	10 min.	20 min.	30 min.	40 min.
Cup 1				
2				
3				
4				

Step 7: Temperature recording chart.

7. Prepare a chart similar to the one illustrated so you can record your observations.

8. Observe and record any change in temperature in any of the four cups every 10 minutes. The rusting process will begin immediately, but the resulting change in temperature will be gradual and small. Make sure that external factors are not affecting the temperature, such as sunlight or heat from a lamp.

Summary of Results

Examine your results and determine whether your hypothesis is correct. Did the temperature rise higher when more wool pads were in the cup? Did it rise in the empty cup? If the reactions resulted in an increase in temperature, then rusting is indeed exothermic. Make sure that your chart shows clearly the result of the tests on each sample.

Change the Variables

You can vary this experiment. Here are some possibilities:

- Other metals will oxidize, though at much slower rates. See if you can measure the temperature change resulting from the oxidation of copper (loops of copper wire may be best).
- Compare the heat energy released by different kinds of oxidation. What about the oxidation you can see occurring on the cut surface of an apple? Find a way to determine if that reaction is exothermic.

Troubleshooter's Guide

Few problems should arise if the steps in this experiment are followed closely. However, when doing experiments involving the mixing of substances, be aware that a number of variables—such as temperature and impurity of substances—can affect your results.

Here is a problem you might encounter:

Problem: You observed little or no temperature change in the cups.

Possible cause: The steel wool is not rusting. Try soaking it in vinegar again for several minutes to remove any protective layers and then repeat the experiment.

Always check first with your teacher before altering the materials used in your experiments.

Experiment 2
Exothermic or Endothermic: Determining whether various chemical reactions are exothermic or endothermic

Purpose/Hypothesis
In this experiment, you will measure the heat energy released or absorbed as four different chemicals (see the materials list) are mixed with water. You expect that the temperature of the solution will go up if the reaction is exothermic and go down if the reaction is endothermic. Before you begin, make an educated guess about the outcome of each reaction based on your knowledge of the chemicals and reactions involved. This educated guess, or prediction, is your **hypothesis.** A hypothesis should explain these things:

- the topic of the experiment
- the variable you will change
- the variable you will measure
- what you expect to happen

What Are the Variables?

Variables are anything that might affect the results of an experiment. Here are the main variables in this experiment:

- the type of reactants used
- the purity of the reactants
- the temperature of the environment in which the samples are tested

In other words, the variables in this experiment are everything that might affect the temperature of the solutions. If you change more than one variable, you will not be able to tell which variable had the most effect on the temperature.

A hypothesis should be brief, specific, and measurable. It must be something you can test through observation. Your experiment will prove or disprove whether your hypothesis is correct. Here is one possible hypothesis for one of the reactions in this experiment: "Mixing water with calcium chloride will produce an exothermic reaction."

In this case, the **variable** you will change is the chemical being reacted with water, and the variable you will measure is the resulting temperature of the solution. In the case of calcium chloride, you expect the temperature to go up.

As a **control experiment,** you will measure the temperature in a beaker of distilled water with no chemical in it. If the temperature changes in the beakers with chemicals as predicted and remains steady in the control beaker, you will know your hypothesis is supported.

Level of Difficulty
Moderate; an adult's supervision is required.

Materials Needed
- 5 glass beakers
- 1 graduated cylinder
- 1 glass stirring rod
- 1 small spoon or spatula

- 1 digital laboratory thermometer
- 1 pint (500 milliliters) distilled water
- 1 tablespoon (14 grams) calcium chloride
- 1 tablespoon (14 grams) sodium hydrocarbonate
- 1 tablespoon (14 grams) ammonium nitrate
- 2 teaspoons (10 milliliters) concentrated sulfuric acid
- safety glasses or goggles
- rubber or surgical gloves

Approximate Budget

$25. (This experiment should be performed only with the appropriate lab equipment and materials. Ask your teacher about ordering the chemicals.)

Timetable

One hour.

Step-by-Step Instructions

1. Place the five beakers on a clean, stable surface and use the graduated cylinder to measure and pour 3 ½ tablespoons (about 50 milliliters) of distilled water into each one.

2. Prepare a chart on which you will record your observations. Your chart should look something like the illustration.

3. Place the thermometer in the first beaker and record the temperature on your chart. This sample, which contains only the distilled water, will be your control.

4. Using the spoon or small spatula, add about half the sample of calcium chloride to

Wear gloves and safety glasses or goggles at all times while performing this experiment.

How to Experiment Safely

This experiment involves dangerous and toxic substances. No part of this experiment should be performed without adult supervision. You must be especially careful handling the sulfuric acid, which is highly corrosive. **Wear gloves and safety glasses or goggles at all times!** When you are finished with the experiment, the chemicals used must be disposed of properly and with supervision. Ask your teacher for help in handling, neutralizing, and disposing of the sulfuric acid.

Step 2: Exothermic vs. endothermic recording chart.

Exothermic vs. Endothermic

	30 sec.	1 min.	1½ min.	2 min.	2½ min.
water only (control)					
water and calcium chloride					
water and sodium hydrocarbonate					
water and ammonium nitrate					
water and sulfuric acid					

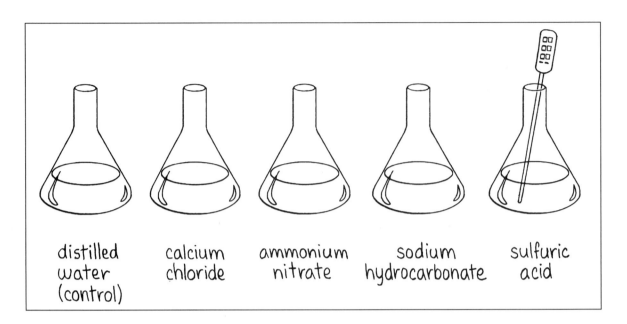

distilled water (control)   calcium chloride   ammonium nitrate   sodium hydrocarbonate   sulfuric acid

Steps 3–7: Exothermic vs. endothermic set-up beakers.

the second beaker. Stir it gently until it is mixed with the distilled water.

5. Place the thermometer in the beaker and note the temperature once each 30 seconds for 5 minutes. Record the temperatures on the chart. When you are done, be sure to rinse the thermometer with room-temperature distilled water.

6. Repeat Steps 4 to 5 for the sodium hydrocarbonate and the ammonium nitrate. Remember to rinse the thermometer, stirring rod, spatula, or spoon in distilled water after each test.

7. In the last beaker, slowly and gently add all of the sulfuric acid to the water. Be careful not to spill or splash the acid. Place the thermometer in the beaker and note the temperature once each 30 seconds for 5 minutes. Record the temperature changes on your chart. When you are done, be sure to rinse the thermometer.

Summary of Results

Examine your results and determine whether each of your hypotheses is correct. If any reactions resulted in an increase in temperature, those reactions are exothermic. If any reactions resulted in a decrease in temperature, they are endothermic. Make sure that your chart shows clearly the result of the tests on each set of reactants. It may be helpful to

Troubleshooter's Guide

When doing experiments involving the mixing of substances, be aware that a number of variables—such as temperature and impurity of substances—can affect your results. When mixing substances, you must keep the mixing containers and utensils clean. Even tiny impurities in a mixture can drastically alter your results.

Here is a problem you might encounter:

Problem: You observed little or no temperature change in the beakers.

Possible cause: You are not placing enough of the solid reactants in the water. Try increasing the amount of solid reactant.

those viewing your results to see a diagram outlining the procedure you followed.

Change the Variables

You can vary this experiment by trying reactions involving different household materials or chemical compounds. Do not mix them with anything other than water. Always check first with your teacher before altering the materials used in your experiments.

Design Your Own Experiment

How to Select a Topic Relating to This Concept

Other kinds of experiments can reveal interesting facts about endothermic and exothermic reactions. Our bodies produce exothermic reactions when we turn food into energy. Can you measure the amount of food energy available in a sample by burning it and measuring the resulting temperature change in a sample of water? Review the description of how cold packs work. Can you think of a way to design a homemade cold pack?

Check the For More Information section and talk with your science teacher or school or community media specialist to start gathering information on chemical reaction questions that interest you.

Steps in the Scientific Method

To do an original experiment, you need to plan carefully and think things through. Otherwise, you might not be sure what question you are answering, what you are or should be measuring, or what your findings prove or disprove.

Here are the steps in designing an experiment:

- State the purpose of—and the underlying question behind—the experiment you propose to do.
- Recognize the variables involved and select one that will help you answer the question at hand.
- State a testable hypothesis, an educated guess about the answer to your question.
- Decide how to change the variable you selected.
- Decide how to measure your results.

Recording Data and Summarizing the Results

In the experiments included here and in any experiments you develop, strive to display your data in accurate and interesting ways. Remember that those who view your results may not have seen the experiment performed, so you must present the information you have gathered as clearly as possible. Including photographs or illustrations of the steps in the experiment is a good way to show a viewer how you got from your hypothesis to your conclusion.

Related Projects

Chemical energy is a basic and crucial part of life processes as well as technological processes. Projects that determine the energy produced by different fuels and compare the by-products of those fuels can help to demonstrate the necessity for developing alternative energy sources. Examining different reactions and determining their endothermic or exothermic rate can help us understand where so much of the energy we use goes.

For More Information

Gillett, Kate, ed. *The Knowledge Factory.* Brookfield, CT: Copper Beech Books, 1996. ❖ Provides some fun and enlightening observations on questions relevant to this topic, along with good ideas for projects and demonstrations.

chapter 6

Chemical Properties

How many ways can you describe a substance? Two common ways are by listing its physical properties and its chemical properties. A **physical property** is a characteristic of a substance that you can detect with your senses, such as its color, shape, size, smell, taste, texture, temperature, density, or volume. For example, a lemon is yellow, oval-shaped, and smaller than a grapefruit. It has a sharp smell and a rough texture.

A **physical change** changes a physical property but does not change the identity or molecular makeup of the substance. One example of a physical change is salt crystals dissolving in water, which changes their shape. When the water evaporates, you can see the salt crystals again, unchanged by being dissolved in the water. Tearing paper into small pieces is also a physical change. The bits of paper look different, but they are still composed of the same molecules as when they were joined together.

A **chemical property** is the ability of a substance to react with other substances or to **decompose.** For example, a chemical property of iron is that it reacts with oxygen and rusts. A chemical property of a substance allows it to undergo a **chemical change.** A chemical change is the change of one or more substances into another substance. A chemical change is also called a **chemical reaction.**

During some chemical reactions, two or more substances are combined to form one new substance. An example is oxygen combining with iron to form rust. This is called a **synthesis reaction.** During other chemical reactions, one substance is broken down into two or more new sub-

Words to Know

Acid:
Substance that when dissolved in water is capable of reacting with a base to form salts and release hydrogen ions.

Base:
Substance that when dissolved in water is capable of reacting with an acid to form salts and release hydrogen ions.

Chemical change:
The change of one or more substances into other substances.

Ice melting is an example of a physical change. (Peter Arnold Inc. Reproduced by permission.)

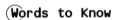
stances. An example of this is hydrogen peroxide, which is used to treat small cuts. It breaks down into oxygen and water in the presence of light, which is why hydrogen peroxide is stored in dark bottles. This is called a **decomposition reaction.** A chemical reaction can be very quick, such as paper burning, or very slow, such as food digesting in your stomach.

What are some examples of chemical properties?

Chemical properties include flammability (the ability to catch on fire), toxicity (the ability to be poisonous), oxidation (the ability to react with oxygen, which causes apple slices to turn brown and iron to rust), radioactivity (spontaneously emitting energy in the form of particles or waves by the disintegration of their atomic nuclei), and sensitivity to light (which causes newspaper to turn yellow).

Being acidic or basic is another kind of chemical property. An **acid** is a substance that can react with, or corrode, other substances. A **base** is a substance that feels slippery when dissolved in water. When an acid and a base are combined, they react chemically with each other to produce new substances: a salt and water.

Many foods contain acids, including tomatoes, lemons, oranges, and carbonated soft drinks. For most people, eating the small amounts of acid in these foods does not cause a problem. In fact, the hydrochloric acid in our stomachs helps produce the chemical reaction called digestion. However, the acid in tomatoes reacts so strongly with aluminum that foods containing tomato sauce should not be stored in aluminum foil. The acid in tomatoes can actually burn holes in the foil.

Acids can also damage the environment. Burning coal produces nitric and sulfuric acids that combine with the water vapor in the air to create acid rain. Acid rain burns trees and plants. It can cause lakes and rivers to become so acidic that fish and plants can no longer survive there.

Burning is a chemical change or reaction, producing new substances. Some substances are more flammable than others. (Photo Researchers Inc. Reproduced by permission.)

Many cleaning products are bases, including soaps, drain cleaners, and ammonia. Basic substances, too, can damage the skin and eyes. For example, some people who breathe ammonia fumes get nosebleeds as the fumes react with the sensitive tissues in their noses.

What happens during a chemical reaction?

In a chemical reaction, the substances you begin with are called **reactants**. The new substances that are formed are called **products**. For example, when the acetic acid in vinegar and baking soda (reactants) are combined, the products are bubbles of carbon dioxide gas, water, and sodium acetate.

(W)ords to Know

Exothermic reaction: A chemical reaction that releases heat or light energy, such as the burning of fuel.

experiment
CENTRAL

experiment
CENTRAL

The chemical properties of the reactants determine what happens during the reaction—and how quickly it happens. For example, one chemical property of magnesium is that it reacts strongly with hydrochloric acid to produce bubbles of hydrogen gas. Not all metals have this property. Dipping a strip of copper into hydrochloric acid produces no hydrogen bubbles. Dipping zinc into the acid results in some bubbles, but fewer than for the magnesium.

In the same way, iron reacts strongly with oxygen to produce rust. However, other metals, such as silver and gold, do not react with oxygen (do not have this chemical property) and so do not rust when exposed to the air.

Many chemical reactions produce energy. For example, when something burns, it produces heat energy. Thus, smoke is one sign of a chemical reaction. Other signs of chemical reactions include foaming, a smell, a sound, and a change in color. A chemical reaction that releases heat or light energy is called an **exothermic reaction.** Examples include fireworks explosions, **luminescent** "light sticks," and the digestive process in your body.

Some chemical reactions absorb heat or light energy and are called **endothermic reactions.** One example is the way green plants absorb sunlight and change it into the chemical energy in sugar and in oxygen.

In the two experiments that follow, you will have an opportunity to produce chemical reactions by using the chemical properties of certain substances. In one experiment, you will combine white glue and borax (a mineral that acts as a laundry booster) to create an entirely new substance. In the second experiment, you will combine water, iodine, and oil to see what kind of chemical reaction occurs. The more you understand about chemical reactions, the better you will understand the workings of the world around—and inside—you.

Experiment 1
Slime: What happens when white glue and borax mix?

Purpose/Hypothesis

In this experiment, you will mix two substances to see if a chemical reaction occurs. The chemical name of one of the substances, white

OPPOSITE PAGE:
The explosion of fireworks produces heat, light, and sound energy in an exothermic reaction. (Photo Researchers Inc. Reproduced by permission.)

Words to Know

Hypothesis:
An idea phrased in the form of a statement that can be tested by observation and/or experiment.

Luminescent:
Producing light through a chemical process.

Physical change:
A change in which the substance keeps its molecular identity, such as a piece of chalk that has been ground up.

Physical property:
A characteristic that you can detect with your senses, such as color and shape.

Product:
A compound that is formed as a result of a chemical reaction.

Reactant:
A compound present at the beginning of a chemical reaction.

glue, is polyvinylacetate. You will mix the polyvinylacetate with borax, a laundry booster (sodium borate). Borax is a natural mineral, found in the ground. It's made of boron, sodium, oxygen, and water. It is used to strengthen the cleaning power of laundry detergents.

To begin the experiment, make an educated guess about what will happen when you combine these two substances. Will there be a chemical reaction? Will it produce a new substance? If so, what might the substance look like? This guess, or prediction, is your **hypothesis.** A hypothesis should explain these things:

- the topic of the experiment
- the **variable** you will change
- the variable you will measure
- what you expect to happen

A hypothesis should be brief, specific, and measurable. It must be something you can test through observation. Your experiment will prove or disprove whether your hypothesis is correct. Here is one possible hypothesis for this experiment: "Mixing polyvinylacetate with borax will create a chemical reaction and produce a new substance."

In this experiment, the variable you will change is the mixing of the two substances, and the variable you will measure (or examine) is the product of this mixture. As a **control experiment**, you will observe a sample of polyvinylacetate that is not mixed with borax to see if a chemical reaction occurs. If only the mixture with the borax in it produces a new substance, your hypothesis will be supported.

Words to Know

Synthesis reaction: A chemical reaction in which two or more substances combine to form a new substance.

Variable: Something that can change the results of an experiment.

Level of Difficulty

Easy/moderate.

Materials Needed

- white glue
- water
- food coloring
- 3 jars with lids
- borax
- labels
- spoons
- measuring spoons
- sealable plastic bag
- goggles

Approximate Budget

Up to $5.

Timetable

10 minutes to set up; 1 hour to observe.

Step-by-Step Instructions

1. Measure 3 tablespoons (44 ml) of water and the same amount of white glue into one jar.

2. Add several drops of food coloring to the jar.

3. Close the jar and shake the mixture vigorously until the glue dissolves in the water. Label the jar "experiment."

4. Repeat Steps 1 to 3, using another jar, and label this jar "control."

5. In the third jar, put 3 tablespoons (44 ml) of water. Slowly pour in 2 tablespoons (30 ml) of borax. Allow the mixture to settle for a minute.

 How to Experiment Safely

Wear goggles to protect your eyes from any splashes. Do NOT taste any mixtures or the product that results from the chemical reaction. Avoid getting the product from this experiment on clothing, carpeting, or furniture, as it might leave a stain.

Step 7: Use a spoon to scrape the wet borax mixture into the experiment jar.

Step 9: Recording table for Experiment 1.

6. Carefully pour the excess water from the third jar down a sink drain.

7. Use a spoon to scrape the wet borax mixture into the experiment jar.

8. With the lids closed, shake both the experiment and control jars for at least two minutes.

9. Record your observations of the experiment jar and the control jar in a table simi-

Table 1: Observations of Change

	Experiment jar	Control jar
Beginning of experiment		
After ½ hour		
After 1 hour		

lar to the one illustrated. Wait half an hour and record them again. After another half an hour, record your final observations.

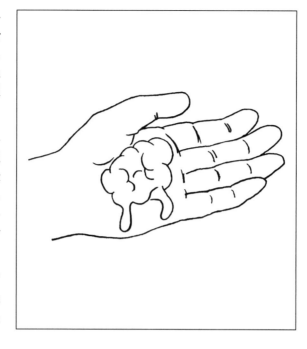

Step 10: Observe and experiment with the "slime" you have created.

10. Open the experimental jar and remove the product you have created. Observe and experiment with its new physical properties.

11. Store your "slime" in the sealable plastic bag to keep it from spoiling.

Summary of Results

Study your observations and decide whether your hypothesis was correct. Did the combination of white glue and borax produce a chemical reaction? How do you know? Did the same reaction occur in the control jar without the borax?

What happened here? In a liquid form, the molecules in polyvinylacetate are separate, allowing the glue to flow. When you added the borax, a chemical reaction caused the molecules in the white glue to wrap around each other, forming a soft ball. The combination of the two substances produced an entirely new substance that looks and feels like slime.

Write a paragraph summarizing your findings and explaining whether they support your hypothesis.

Change the Variables

You can vary this experiment by changing the amount of borax you mix with the white glue solution. Your products will range from sticky slime, to a bouncy ball, to a very hard ball.

You might also experiment with other types of glue, such as gel glue and washable glue, to see if they form the same kind of product when mixed with borax.

Experiment 2
Chemical Reactions: What happens when mineral oil, water, and iodine mix?

Purpose/Hypothesis

In this experiment, you will mix water with iodine and then add mineral oil to see whether a chemical reaction occurs. Remember the possible signs of a chemical reaction: the production or absorption of heat or light energy, smoke, bubbles of gas, a smell, a sound, and a change in color.

You know that water and oil do not mix. Instead, they remain as separate layers. You probably also know that a combination of water and oil does not produce any sign of a chemical reaction. If such a reaction is to occur, it must be caused by the iodine. Make an educated guess about the outcome of this experiment. This guess, or prediction, is your **hypothesis.** A hypothesis should explain these things:

- the topic of the experiment
- the **variable** you will change
- the variable you will measure
- what you expect to happen

What Are the Variables?

Variables are anything that might affect the results of an experiment. Here are the main variables in this experiment:

- the amount of iodine added to the water in the experiment

- the temperature of the ingredients (they will remain at room temperature to control this variable)

- the kind of oil used (other kinds of oil may react differently)

   If you change more than one variable, you will not be able to tell which variable had the most effect on the chemical reaction.

A hypothesis should be brief, specific, and measurable. It must be something you can test through observation. Your experiment will prove or disprove whether your hypothesis is correct. Here is one possible hypothesis for this experiment: "Iodine will cause a chemical reaction when mixed with mineral oil and water."

In this case, the variable you will change is the presence of iodine. The variable you will measure or observe is evidence of a chemical reaction. As your control experiment, you will combine mineral oil and water, without adding iodine, and watch for signs of a chemical reaction. If a change occurs only in the mixture with the iodine, your hypothesis will be supported.

Level of Difficulty

Easy/moderate.

Materials Needed

Note: All ingredients should be at room temperature.

- two jars with lids, such as peanut butter jars
- labels
- water
- a container of iodine with a dropper
- mineral oil
- measuring cups
- goggles

Approximate Budget

$5 for iodine and mineral oil; other materials should be available in the average household.

Timetable

30 minutes.

How to Experiment Safely

Wear goggles to protect your eyes from possible splashes of iodine. Avoid getting iodine or mineral oil on your clothing or furniture, as it will stain.

Step 3: Add about 5 drops of iodine to the experiment jar.

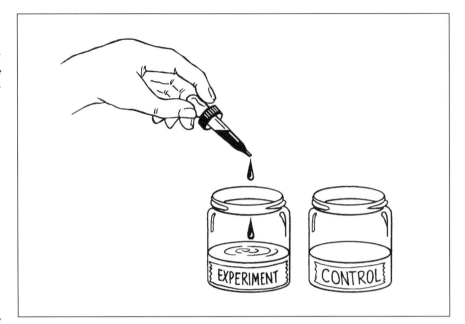

Step 4: Recording table for Experiment 2.

Table 2: Evidence of a Chemical Change

	Experiment jar	Control jar
After adding iodine		————————
After adding mineral oil		
After vigorous shaking		

Step-by-Step Instructions

1. Label one jar "experiment" and one jar "control."

2. Pour 1/4 cup (60 ml) of water into each jar.

3. Add about 5 drops of iodine to the experiment jar.

4. Record your observations on a table similar to the one illustrated.

5. Pour 1/4 cup (60 ml) of mineral oil into each jar. Record your observations in the table.

6. Shake both jars, one in each hand, for two minutes. Again, record any changes you observe.

Summary of Results

Study the observations on your table and decide whether your hypothesis was correct. Did a chemical reaction take place in the mixture containing iodine? How can you tell? Did a chemical reaction occur in the mixture without the iodine? Write a paragraph summarizing your findings and explaining whether they support your hypothesis.

When you shook the mixture containing iodine, the iodine moved from the water into the oil, causing a color change, which is evidence of a chemical reaction. If you shake the experiment jar long enough, all the iodine will move into the oil, and the water will become clear again. The iodine causes the chemical reaction, so the mixture without iodine did not change.

Change the Variables

Here are some ways you can vary this experiment:

- Use other kinds of oil, such as safflower or peanut oil, to see if the same color change results.

Step 6: Shake both jars, one in each hand, for two minutes.

Troubleshooter's Guide

Below is a problem that may arise during this experiment, a possible cause, and a way to remedy the problem.

Problem: The mixture with iodine did not change color.

Possible cause: You did not shake it long enough. Shake it some more and observe what happens.

- Use very hot or icy-cold water to see how a change in temperature affects this chemical reaction.

 # Design Your Own Experiment

How to Select a Topic Relating to this Concept

The world—and your own life—depend on chemical properties and the chemical reactions that result from them. Consider what you would like to know about these properties and reactions. For example, what chemical reactions occur inside your body? Which ones are essential in manufacturing? What chemical reactions help shape the landscape?

Check the For More Information section and talk with your science teacher or school or community media specialist to start gathering information on questions that interest you. As you consider possible experiments, be sure to discuss them with your science teacher or another knowledgeable adult before trying them. Combining certain materials can be dangerous.

Steps in the Scientific Method

To do an original experiment, you need to plan carefully and think things through. Otherwise, you might not be sure which question you are answering, what you are or should be measuring, or what your findings prove or disprove.

Here are the steps in designing an experiment:

- State the purpose of—and the underlying question behind—the experiment you propose to do.
- Recognize the variables involved, and select one that will help you answer the question at hand.
- State a testable hypothesis, an educated guess about the answer to your question.
- Decide how to change the variable you selected.
- Decide how to measure your results.

Recording Data and Summarizing the Results

In your "slime" and iodine experiments, your raw data might include tables, drawings, and photographs of the changes you observed. If you display your experiment, make clear the question you are trying to answer, the variable you changed, the variable you measured, the results, and your conclusions. Explain what materials you used, how long each step took, and other basic information.

Related Projects

You can undertake a variety of projects related to chemical reactions. For example, a number of chemical reactions occur in the kitchen as food cooks on the stove or bakes in the oven. Breads and cakes rise because of a chemical reaction. Some medicines for an upset stomach depend on chemical reactions to cause fizz in a glass of water. You can even make pennies turn green because of a chemical reaction!

For More Information

Gardner, Robert. *Science Projects about Chemistry*. Hillside, NJ: Enslow Publishers, 1994. ❖ Focuses on experiments in causing and analyzing chemical reactions.

Mebane, Robert, and Thomas Rybolt. *Adventures with Atoms and Molecules*. Hillside, NJ: Enslow Publishers, 1991. ❖ Clearly describes 30 doable experiments in chemistry and chemical reactions.

VanCleave, Janice. *A+ Projects in Chemistry*. New York: Wiley and Sons, 1993. ❖ Outlines experiments that show chemical reactions relating to the weather, biochemistry, electricity, and other topics.

chapter 7

Chlorophyll

Chlorophyll is the green pigment that gives leaves their color. Acting as a solar collector, chlorophyll absorbs light energy from the sun and traps it. This trapped energy is stored, then used to form sugar and oxygen out of carbon dioxide from the air and water from the soil. This extraordinary process is called **photosynthesis.** It is the way a plant makes its own food. But the key to this process is chlorophyll.

What's this green thing?

Pierre Joseph Pelletier and Joseph Biernaime Caventou were French chemists who worked together in the early nineteenth century in a new field called **pharmacology,** the science of preparing medical drugs. These chemists would later discover quinine, caffeine, and other specialized plant products. In 1817, however, they isolated an important plant substance they called chlorophyll, from the Greek words meaning "green leaf." Scientists first thought that chlorophyll was distributed throughout plant cells. But in 1865 the German botanist Julius von Sachs discovered that this pigment is found within sacs called **chloroplasts.** Chlorophyll molecules are arranged in clusters within these chloroplasts.

One-celled plants, such as algae, contain chlorophyll. They live in water, growing near the surface and the light, or on moist surfaces. Multicelled plants—usually land plants such as mosses, ferns, and seed plants—have chlorophyll-loaded chloroplasts in their stems and leaves. These plants all need light to activate the chlorophyll. Plants such as algae require low light, and certain land plants, such as philodendron,

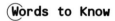

Words to Know

Abscission:
Barrier of special cells created at the base of leaves in autumn.

Anthocyanin:
Red pigment found in leaves, petals, stems, and other parts of a plant.

Carbohydrate:
Any of several compounds composed of carbon, hydrogen, and oxygen, which are used as food for plants and animals.

Carotene:
Yellowish-orange pigment present in most leaves.

experiment
CENTRAL

LEFT: *Chlorophyll clusters in the leaves of this healthy rhododendron plant trap solar energy. (Photo Researchers Inc. Reproduced by permission.)*

RIGHT: *An unhealthy rhododendron plant. If plants do not get enough light to activate their chlorophyll clusters, they cannot make enough food to survive. (Photo Researchers Inc. Reproduced by permission.)*

survive well in low levels of sunlight also. Some houseplants thrive in artificial light, while other plants require high levels of sunlight.

Why leaves change color

Pigments are substances that appear colored to the human eye because of the wavelengths of light they reflect. A pigment absorbs all other wavelengths of light and only reflects the wavelength that we see as a color. For example, a green pigment, like chlorophyll, absorbs all wavelengths except green. Many different pigments are present in sacs within the plant cell. There are two related chlorophyll pigments, chlorophyll A and chlorophyll B. Both appear green, with just a slight color variation from each other. **Carotene,** a yellowish-orange pigment, and **xanthophyll,** a yellow pigment, are also present in most leaves. Some plants have a red color in their petals, stems, and leaves called **anthocyanin.** The different pigments in a plant allow the plant to absorb different light wavelengths. Overall, the greenish chlorophyll pigment is the one that is most plentiful. It is considered a primary pigment, and the secondary pigments act as a support team to help the plant absorb more light energy.

Deciduous trees shed their leaves in the autumn. The joining place where the leaf meets the twig is called an **abscission.** The first step in

Words to Know

Chloroplasts:
Small structures in plant cells that contain chlorophyll and in which the process of photosynthesis takes place.

Chromatography:
A method for identifying the components of a substance based on their characteristic colors.

Control experiment:
A set-up that is identical to the experiment but is not affected by the variable that affects the experimental group.

Germination:
First stage in development of a plant seed.

Hypothesis:
An idea in the form of a statement that can be tested by observation and/or experiment.

Pharmacology:
The science dealing with the properties, reactions, and therapeutic values of drugs.

the process that causes leaves to fall occurs when cork cells develop under the abscission. This cork layer blocks nutrients that travel to and from the leaf. Then the leaf begins to die.

Because chlorophyll breaks down faster than the other pigments, the green leaves begin their gradual color change. As the chlorophyll decomposes, the yellow and orange colors from the carotene and xanthophyll stand out. Trees with anthocyanin pigments show bright red leaves in the fall. Anthocyanin pigments need high light intensity and sugar content for their formation, so fiery red leaves usually emerge after bright autumn days. Cool nights act as a refrigerator, preserving the sugar in the leaves.

Chlorophyll and other pigments are unique in their function as food makers. Uncovering their presence in plants through experiments will help you "see" them.

Experiment 1
Plant Pigments: Can pigments be separated?

Purpose/Hypothesis

In this experiment you will discover what pigments are present in various plants using **chromatography,** an identification technique based on

What Are the Variables?

Variables are anything that might affect the results of an experiment. Here are the main variables in this experiment:

- the type and part of the plant being used (Example: carrot roots contain mainly carotene; carrot leaves contain mainly chlorophyll.)

- the season in which the plant was harvested (Example: if the plant was harvested in the spring, the leaves contain abundant chlorophyll; in the fall, the leaves have more carotene, xanthophyll, and anthocyanin.)

- the maturity of the specimen (Example: leaves from the heart of a celery plant are yellow from xanthophyll; as leaves mature, chlorophyll builds up.)

In other words, the variables in this experiment are everything that might affect the colors you find. If you change more than one variable, you will not be able to tell which variable had the most effect on the color.

Words to Know

Photosynthesis:
Chemical process by which plants containing chlorophyll use sunlight to manufacture their own food by converting carbon dioxide and water to carbohydrates, releasing oxygen as a by-product.

Pigment:
A substance that displays a color because of the wavelengths of light that it reflects.

Variable:
Something that can change the results of an experiment.

Wavelength:
The peak-to-peak distance between successive waves. Red has the longest wavelength of all visible light, and violet has the shortest wavelength.

Xanthophyll:
Yellow pigment found in leaves.

color. You will cut up various plants and boil them in water, then add a small amount of alcohol to help release the pigments from the plants.

To begin the experiment, use what you know about chlorophyll and other pigments found in plants to make an educated guess about what colors you will find. This educated guess, or prediction, is your **hypothesis.** A hypothesis should explain these things:

- the topic of the experiment
- the **variable** you will change
- the variable you will measure
- what you expect to happen

A hypothesis should be brief, specific, and measurable. It must be something you can test through observation. Your experiment will prove or disprove whether your hypothesis is correct. Here is one possible hypothesis for this experiment: "Primary pigments, such as blue-green chlorophyll, and secondary pigments, such as yellow-orange carotene, yellow xanthophyll, and red anthocyanin, are all present in leaves."

In this case, the variable you will change is the type and part of the plant being tested, and the variable you will measure is the resulting mix of colors. A bowl filled with various food colorings will serve as a **control experiment** to allow you measure the effectiveness of the color separation method. If you find many different colors present in your experimental solutions, you will know your hypothesis is correct.

Level of Difficulty
Moderate.

Materials Needed
- 1 cup (236 milliliters) of spinach leaves, cut up
- 1 cup (236 milliliters) of parsley leaves, cut up
- 1 cup (236 milliliters) of coleus leaves (houseplant with variegated leaves), cut up
- food coloring (red, blue, and yellow)
- filter paper (strong paper towels also will work)
- rubbing alcohol 70 percent
- 4 bowls
- 4 glass cups or beakers
- cooking pot
- labels
- 4 paper clips
- measuring spoons and cups
- water
- goggles

Approximate Budget
$10 for the fresh parsley, spinach, and a coleus plant.

Timetable
Approximately 2 hours.

Step-by-Step Instructions
1. Place one cup of water in a pot and bring it to a boil. Add 20 drops of each color of food coloring and boil for 10 minutes more. Remove the pot from stove and allow to cool. Pour the solution into a bowl and add 4 tablespoons of alcohol. Label the bowl "#1." This will be your control solution.

Step 6: Four cups with pigment solutions (control, spinach, parsley, and coleus).

food coloring control

spinach solution

parsley solution

coleus solution

#1 control

#2 spinach

#3 parsley

#4 coleus

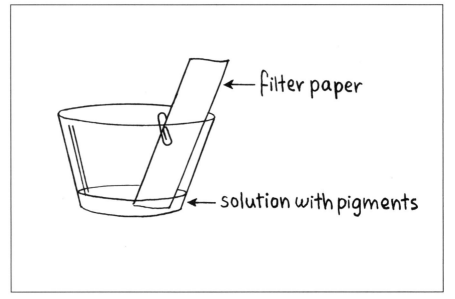

Step 7: Filter paper strip in cup, held in place with a paper clip.

← filter paper

← solution with pigments

2. Wash the pot and add one cup of water and bring it to a boil. Add the cut-up spinach leaves. Boil for 10 minutes more. Remove the pot from stove and allow to cool. Pour the solution into another bowl and add 4 tablespoons of alcohol. Label the bowl "#2."

3. Repeat Step 2, substituting parsley for spinach. Label bowl "#3."

4. Repeat Step 2 again, substituting coleus leaves. Label bowl "#4."

How to Experiment Safely

This experiment requires the use of a stove or bunsen burner to boil the solutions. Use caution when cooking the solution and ask an adult for assistance. When handling alcohol, wear goggles and be careful not to spill it on your skin or in your eyes. Keep alcohol away from the stove or open flame.

5. Cut the filter paper into 1-inch-wide (2.5-centimeter) strips. These will be your chromatography papers.

6. Label the cups #1, #2, #3, and #4. Now pour 0.25 inch (0.6 centimeter) of the liquid solution from each bowl into the appropriate numbered cup.

7. Place a filter paper strip into each cup as illustrated. Use a paper clip to hold the paper to the cup. Make sure only the bottom of the filter paper touches the solution.

8. Leave the experiment undisturbed for 30 to 60 minutes. Notice how the solution creeps up the filter paper.

9. Stop the experiment when a pigment reaches the top of the filter paper. Place the pieces of paper on a clean, flat surface to dry.

Sample diagram of chromatography paper from one of the solutions.

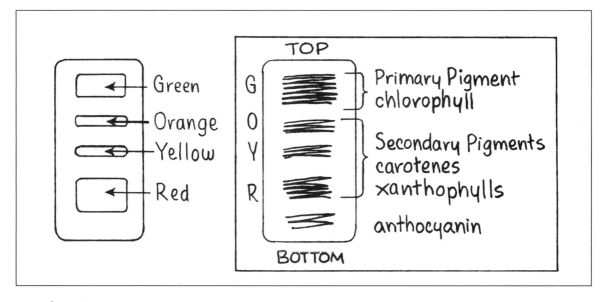

Troubleshooter's Guide

Here are some problems that may occur in this experiment, some possible causes, and ways to remedy the problems:

Problem: The pigment does not go up the paper.

Possible cause: The paper is wet. Make sure the paper is thoroughly dry before inserting it in the solution. Also make sure the paper is touching the solution.

Problem: The control experiment worked well, but the spinach, parsley, and coleus solutions are very light.

Possible cause: The solutions are too weak. Place more leaves into the pot and boil the solution longer. Use a low flame, and be cautious when reheating as the mixture contains alcohol.

Summary of Results

Make a diagram recording what colors appeared on your chromatography papers (see sample diagram). The pigments may fade over time, so record the results the same day.

Reflect on your original hypothesis. Were you able to detect the primary and secondary pigments present in all the leaves? Were pigments present in your control experiment? Which plant(s) contained the most secondary pigments? Which contained the most primary pigments?

Experiment 2
Response to Light: Do plants grow differently in different colors of light?

Purpose/Hypothesis

In this experiment you will test the growth of plant seedlings under different colors of light. Within the cells of a plant's leaves and stems, there are various pigments that react to light to perform photosynthesis. The pigments vary in color and concentration. Each pigment absorbs all colors of light except the color of the pigment itself, which is reflected. For example, if a plant contains mostly green pigments

What Are the Variables?

Variables are anything that might affect the results of an experiment. Here are the main variables in this experiment:

- the type of seedlings being used
- the strength of light (wattage)
- the wavelengths (colors) of light being tested
- the amount of water given to the seedlings

In other words, the variables in this experiment are everything that might affect the growth of the seedlings. If you change more than one variable, you will not be able to tell which variable most affected the seedlings' growth.

such as chlorophyll, the plant should grow well under all colors of light except green because it reflects most of the green light without absorbing it. As a result, the plant is "starved" for light and cannot perform the photosynthesis process needed to produce food and grow.

To begin this experiment, use what you know about chlorophyll and the pigment colors found in plants to make an educated guess about how plants will grow under various colors of light. This educated guess, or prediction, is your **hypothesis.** A hypothesis should explain these things:

- the topic of the experiment
- the **variable** you will change
- the variable you will measure
- what you expect to happen

A hypothesis should be brief, specific, and measurable. It must be something you can test through observation. Your experiment will prove or disprove whether your hypothesis is correct. Here is one possible hypothesis for this experiment: "Seedlings will grow best under white light, because they can absorb more energy from the wide range of wavelengths present. They will grow worst under green light, because that is the color of the dominant pigment contained in their leaves and stems, and most of that light will be reflected instead of absorbed."

In this case, the variable you will change is the color of the light, and the variable you will measure is the amount of growth of the seedlings over a period of several weeks. If the seedlings grow best under white light and worst under green light, you will know your hypothesis is correct.

Level of Difficulty
Moderate. (However, great care of the seedlings must be taken to ensure their growth.)

Materials Needed
- 4 boxes, 24 inches (60 centimeters) square in size, open on one side
- aluminum foil
- 4 light fixtures with 40-watt white incandescent bulbs, such as small desk lamps
- 4 plastic filters about 12 inches (30 centimeters) square, such as cellophane in clear, green, blue, and red
- black plastic cut from a garbage bag
- 4 shallow trays filled with potting soil
- 40 bean seeds, such as lima, kidney, or others. (Use all of one type.)
- water

Note: If you are unable to get light fixtures to use, use natural sunlight and modify the setup described in the following procedure.

Approximate Budget
$30-$35 for light fixtures, if necessary, and $5 for seeds and cellophane.

Timetable
Approximately 2 months—about 20 days for the seeds to **germinate,** and 2 to 3 weeks before the first true leaves appear.

How to Experiment Safely

Incandescent light fixtures and bulbs can get hot. Do not handle or leave the lights on for more than 10 hours at a time. Never leave them on overnight. Keep them a safe distance from the cellophane filters at all times.

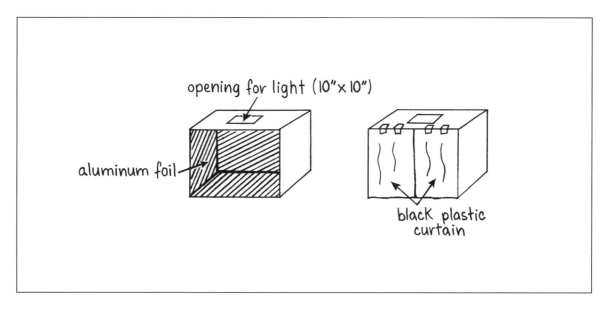

opening for light (10" x 10")

aluminum foil

black plastic curtain

Step-by-Step Instructions

Step 1: Set-up of boxes with aluminum foil and black plastic.

1. Set up four identical boxes. Line the inside of each box with aluminum foil. Cover the front opening with black plastic. Cut a hole in the top, about 10 inches by 10 inches, (25 centimeters by 25 centimeters), to allow light to enter.

2. Tape a different color plastic filter over the hole on each box.

3. Position a light fixture approximately 12 inches (30 centimeters) above the opening on each box and aim the light inside the box.

4. Place a tray of soil into each box and plant 10 seeds slightly below the surface of the soil. Water gently.

Step 3: Light fixture over opening of box.

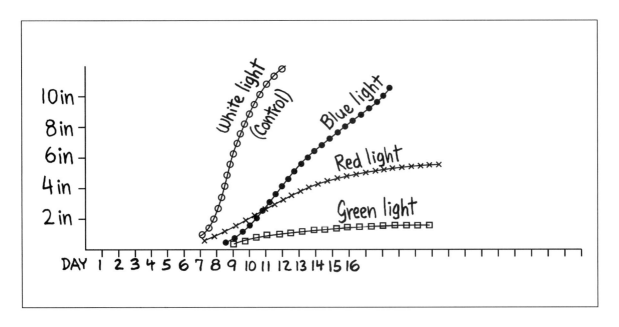

Step 6: Sample seed growth recording chart.

5. Turn the lights on for 8 to 10 hours a day. Monitor the soil moisture and water gently when needed.

6. Record the seed growth in each box. Record which seedling is the tallest daily for one month after the seeds sprout or until the seedlings reaches the filter.

Summary of Results

Make a chart to track the growth of the seedlings. Reflect on your hypothesis. Were the seedlings more responsive to one color of light? What color stimulated growth the least? Is that color the seedlings' most dominant pigment? Summarize your results in writing.

 Troubleshooter's Guide

Here is a problem that may arise in this experiment, a possible cause, and a way to remedy it:

Problem: The seeds did not grow.

Possible Cause: The seeds might be too old. You can try again with new seeds or accept the results if you think it was the lighting. If they died from not getting enough water, then try again.

experiment
CENTRAL

 **Design Your Own Experiment**

How to Select a Topic Relating to This Concept

All the colors in plants and animals are due to pigments, which have many functions. Chlorophyll's function is producing energy for photosynthesis. Melanin is a skin pigment that protects people and animals from harmful solar radiation.

Check the For More Information section and talk with your science teacher or school or community media specialist to start gathering information on questions that interest you about chlorophyll and other pigments. As you consider possible experiments, be sure to discuss them with your science teacher or another knowledgeable adult before trying them. Some pigments might be dangerous.

Steps in the Scientific Method

To do an original experiment, you need to plan carefully and think things through. Otherwise, you might not be sure what questions you're answering, what you are or should be measuring, or what your findings prove or disprove.

Here are the steps in designing an experiment:

- State the purpose of—and underlying question behind—the experiment you propose to do.
- Recognize the variables involved, and select one that will help you answer the question at hand.
- State a testable hypothesis, an educated guess about the answer to your question.
- Decide how to change the variable you selected.
- Decide how to measure your results.

Recording Data and Summarizing Results

Think of how you can share your results with others. Charts, graphs, and diagrams of the progress and results of the experiments are very helpful in informing others about an experiment.

Related Projects

You can create an experiment on pigments by discovering how to extract pigments from their source in nature. Or you could take an extracted pigment and find a use for it. For example, purple grape juice can be used as an acid/base indicator.

For More Information

Halpern, Robert. *Green Planet Rescue.* New York: Franklin Watts, 1993. ❖ Discusses the importance of plants and what can be done to protect plants that face extinction.

Kalman, Bobbie. *How A Plant Grows.* New York: Crabtree Publishing, 1997. ❖ Examines the stages of a seed plant's development and includes activities on how to grow plants.

chapter 8

Composting/Landfills

Composting is the process in which **organic** wastes are broken down biologically and become dark, fertile soil called **humus.** An ancient practice, composting probably began when the original hunter-gatherers began cultivating food and saw that crops grew better in areas where the soil contained **manure,** the waste matter of animals.

Agricultural composting with manure was being used in the Mesopotamia Valley in Asia as early as 13 B.C. Not surprisingly, Native American tribes practiced composting long ago, as did the first colonists who arrived in North America.

A smelly solution

French chemist Jean Baptiste Boussingault (1802–1887) made significant contributions to agricultural chemistry by suggesting that good soil was made by the action of **microorganisms,** bacteria, and fungi that break down waste. Working on his farm, he applied and studied the results of organic methods of farming from 1834 to 1876.

At that time, composting used mostly animal manure or dead fish, as well as nutrient-rich muck from swampy areas. By the twentieth century, large animals such as the buffalo, whose droppings fertilized the prairie soil, were disappearing as were many of the farming communities that contributed barnyard manure to compost piles.

In 1934, Sir Albert Howard, an Englishman, developed the modern organic concept of farming. Through several years of research in Indore, India, he formulated the Indore method, a process that used

Words to Know

Aerobic:
A process that requires oxygen.

Anaerobic:
A process that does not require oxygen.

Biodegradable:
Materials that can undergo decomposition by biological variables.

Biological variables:
Living factors such as bacteria, fungi, and animals that can affect the processes that occur in nature and in an experiment.

Backyard compost bins are simple to use. (Peter Arnold Inc. Reproduced by permission.)

three times more plant waste than manure in sandwich-like layers of green or wet material. Howard also pointed out the importance of microorganisms in the process. In 1942, J.I. Rodale began publishing *Organic Farming and Gardening.* Rodale used Howard's techniques and experimented with his own. He is considered the pioneer of organic methods of farming in the United States.

Chomping microbes

How does composting work? Let us begin with the basics, the organic waste. That would be vegetable scraps such as carrot tops and peelings, plus leaves, paper bags, grass clippings, tea bags, and coffee grounds. Carbon in these organic waste materials provides food for the microorganisms, starting the composting process. When these microbes chomp away and begin digesting, the carbon is burned off or oxidized, causing the composting pile to heat up. The heat kills any harmful organisms. **Macroorganisms**—such as earthworms, insects, mites, and

Macroorganisms, such as earthworms, chew organic matter into smaller pieces. (Photo Researchers Inc. Reproduced by permission.)

grubs—continue the composting process by chewing the organic matter into smaller pieces. Through digestion and excretion, both types of organisms release important chemicals into the compost mass, which then becomes humus, a nutrient-rich soil.

The transformation is speeded up by a balanced supply of carbon and nitrogen, the oxygen required by the microorganisms, enough moisture to allow biological activity, and suitable temperatures. But it is really the diverse microorganisms that chomp away and activate the process. Without them, we would be buried in wastes.

In the United States, more garbage is generated than in any other country in the world. Materials that could be used in composting make up 20 to 30 percent of the **waste stream**—the waste output of any area or facility. This figure doubles in the autumn when leaves and garden clippings are added. All this waste winds up in **landfills.**

Landfills that raised the roof

Landfills are huge depressions in the ground or equally huge mounds above ground where garbage is dumped. Like compost piles, landfills also have centuries-old beginnings. The ancient cities of the Middle East were built up over time on mounds that contained the remains of everyday life. In excavations of the ancient city of Troy, in what is now Greece, building floors were found to have layers of animal bones and

The Fresh Kills landfill in Staten Island, New York, is one of many that will soon be full. (The Stock Market. Reproduced by permission.)

artifacts that had been alternated with layers of clay. These layers piled up until it was necessary to raise roofs and rebuild doorways.

During the Bronze Age (3000–1000 B.C.), the city of Troy rose about 4.7 feet (1.4 meters) each century (100 years) because of these accumulations. Landfilling has also been used to extend shorelines. In New York City during the eighteenth century, shorefront roads were extended into the water by landfill that included broken dishes, old shoes, and even the rotted hulls of boats.

Sanitary landfills

In the 1930s, solid waste materials covered with soil became known as "sanitary landfill." As with composting, a **decomposition** process takes place in landfills. The process has an aerobic and an anaerobic phase. **Aerobic** means requiring oxygen. **Anaerobic** means functioning without oxygen. In the aerobic phase, **biodegradable** solid wastes react with the landfill's oxygen to form carbon dioxide and water. The landfill temperature rises and a weak acid forms within the water, dissolving some of the minerals. Microorganisms that do not need oxygen break down wastes into hydrogen, ammonia, carbon dioxide, and inorganic acids during the anaerobic stage. Gas in the form of carbon monoxide and methane is produced in the third stage of decomposition.

In a landfill, many of the materials, such as plastic, glass, and aluminum cans, containers, and bottles, can take up to forty years or more to decompose. As a result, these materials are quickly filling the space available in landfills. That is why **recycling** is encouraged in most communities. In recycling, waste materials are used to produce new materials.

In the United States, about 80 percent of the garbage is dumped into landfills. Landfills are not bottomless pits. Soon half of all U.S. landfills will be full. In New York State, for example, all landfills will be officially closed by the end of 2000. Only Fresh Kills, the largest sanitary landfill in the world, still operates, receiving more than 200 tons (180 metric tons) of waste a day. Understanding how composting and landfills work helps everyone become more aware of what happens to the garbage that is thrown away.

Experiment 1
Living Landfill: What effect do the microorganisms in soil have on the decomposition process?

Purpose/Hypothesis
The purpose of this experiment is to determine what happens to common household items that are discarded and placed in a landfill. In nature, physical, chemical, and biological factors act upon our waste and work together in the process of decomposition. This experiment will determine what action organisms in the soil have on garbage. Before you begin, make an educated guess about the outcome of this experiment based on your knowledge of composting and decomposition. This educated guess, or prediction, is your **hypothesis**. A hypothesis should explain these things:

- the topic of the experiment
- the variable you will change
- the variable you will measure
- what you expect to happen

A hypothesis should be brief, specific, and measurable. It must be something you can test through observation. Your experiment will prove or disprove whether your hypothesis is correct. Here is one possible hypothesis for this experiment: "Household garbage covered with soil will decay faster than garbage not covered with soil."

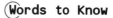

Words to Know

Variable:
Something that can affect the results of an experiment.

Waste stream:
The waste materials generated by the population of an area, or by a specific industrial process, and removed for disposal.

What Are the Variables?

Variables are anything that might affect the results of an experiment. This experiment involves both environmental variables and biological variables. Here are the main variables in this experiment:

- the presence of air—needed for living things, bacteria, fungi, etc.
- the presence and amount of water—also needed for living things, bacteria, fungi, etc.
- the temperature—warm temperatures promote biological decomposition; cold temperatures (especially freezing temperatures) can cause physical breakdown when water freezes and expands.
- the pH—extreme pH levels can stop biological activity and cause chemical breakdown. For example, strong acids and bases are corrosive and can chemically break down debris.
- the amount and types of bacteria present— these microscopic organisms in the soil consume organic matter
- the amount and types of fungi—these microscopic and macroscopic organisms also consume organic matter

In other words, the variables in this experiment are everything that might affect the amount of decomposition of the garbage. If you change more than one variable, you will not be able to tell which variable had the most effect on the decomposition.

In this case, the variable you will change is the presence or absence of soil, and the variable you will measure is the differences in condition between the garbage in the two bags after two to three months. If the garbage in the bag with soil has decayed more than the garbage in the bag without soil, you will know your hypothesis is correct.

Level of Difficulty

Easy/Moderate, because of the time involved.

Materials Needed

- Two 1-gallon plastic bags with holes. Each bag should have approximately 20 randomly placed holes. The holes should be about 0.5 inch (1.25 centimeters) in diameter. A hole puncher or pencil can accomplish this task.
- 2 twist ties to seal bags
- 5 pairs of household garbage items (for example, 2 food containers, 2 glass bottles, 2 pieces of leftover food or bones, 2 small sticks or leaves, and 2 metal cans)
- permanent marker
- 3 to 5 cups of soil
- plastic gloves

Approximate Budget

$5 for the materials that cannot be found in your household or at school.

Timetable

Three to four months for decomposition to take place.

Step-by-Step Instructions

1. Prepare a sketch and written description of the materials being placed into each bag.

Materials needed for Experiment 1.

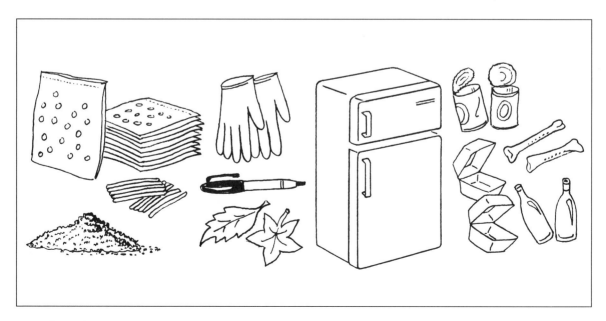

How to Experiment Safely

Always wear gloves when handling garbage. Use caution when handling sharp objects, glass, or metal.

2. Prepare the control experiment. The control for this experiment will remove as many variables as possible from the test in order to see the results from a single variable. In one bag place one of each item and sprinkle a little water over them. Do not add soil to the control bag. Seal the bag with a twist tie.

3. Prepare the test bag. In the other bag, place one of each item. Add to the bag 3 to 5 cups of soil to cover the garbage. Sprinkle the mixture with water and seal the bag with a twist tie.

4. Label each bag ("control" or "test") and place both of them outside in a shady spot.

5. Open the bags every 2 to 3 weeks, sprinkle more water over the contents, and reseal the bags.

Step 4: Completed control and test bags.

OPPOSITE PAGE:
Sample landfills results chart for Experiment 1.

	Description	Before sketch	After sketch
CONTROL ONE			
TEST ONE			
CONTROL TWO			
TEST TWO			
CONTROL THREE			
TEST THREE			

Troubleshooter's Guide

Because this experiment requires living organisms to act upon waste, it is essential that the conditions in the landfill be correct. Factors such as extreme weather conditions or excessive temperatures could cause undesirable results in your experiment. If you should have problems, try the following tips: Always keep soil moist, not wet. Make sure the soil does not get too hot or cold. Temperatures between 40°F and 100°F (4°C and 38°C) are ideal. If you use black garbage bags, keep them out of the sun, because the dark color absorbs light and can overheat the soil easily.

6. After 3 months, open the bags and pour out the contents of each onto separate pieces of newspaper. Remember to wear gloves. Record what changes have occurred to each item. Compare the differences in breakdown between the control and test bags.

Summary of Results

When analyzing the contents of each bag, sketch the objects and write a brief description of their conditions. Look for any activity of organisms like worms or insects. If anything is smelly, slimy, or has a black stain due to bacterial action, record it in the result chart (see sample chart). Note the difference in decay between the organic waste (food) and the inorganic waste (containers).

Change the Variables

You can vary this experiment by changing the variables. For example, you can place one bag in a chilly basement or the freezer and the other bag in a warm spot outside to determine the effect of temperature. You could also add water to one bag, but not to the other, to determine the effect of water. To determine the effect of pH on decomposition, you could add an acidic material like vinegar to one bag, and add water to the other bag.

Experiment 2
Composting: Using organic material to grow plants

Purpose/Hypothesis

This experiment will examine the principle of composting, the process of converting complex organic matter into the basic nutrients needed by living organisms. This experiment will utilize organic waste (house-

What Are the Variables?

Variables are anything that might affect the results of an experiment. Here are the main variables in this experiment:

- the presence of air—needed for living things, bacteria, fungi, etc.

- the presence and amount of water—also needed for living things, bacteria, fungi, etc.

- the temperature—warm temperatures promote biological decomposition; cold temperatures (especially freezing temperatures) can cause physical breakdown when water freezes and expands.

- the pH—extreme pH levels can stop biological activity and cause chemical breakdown. For example, strong acids and bases are corrosive and can chemically break down debris.

- the amount and types of bacteria present— these microscopic organisms in the soil consume organic matter

- the amount and types of fungi—these microscopic and macroscopic organisms also consume organic matter

- the type of plant—roots of plants aid in the physical breakdown of material by helping to separate materials as the roots grow through the waste

In other words, the variables in this experiment are everything that might affect the degree of decomposition and the growth rate of the plants. If you change more than one variable, you will not be able to tell which variable had the most effect on the decomposition and plant growth.

hold and yard waste) as nutrients for plants. It will allow you to investigate which waste products can be composted and best utilized by plants. Before you begin, make an educated guess about the outcome of the experiment based on your knowledge of composting and decomposition. This educated guess, or prediction, is your **hypothesis**. A hypothesis should explain these things:

- the topic of the experiment
- the variable you will change
- the variable you will measure
- what you expect to happen

A hypothesis should be brief, specific, and measurable. It must be something you can test through observation. Your experiment will prove or disprove whether your hypothesis is correct. Here is one possible hypothesis for this experiment: "Yard waste will break down faster than household waste and will provide more nutrients for plants."

In this case, the **variable** you will change is the type of waste used to make compost, either yard waste or household waste, and the variable you will measure is the amount of decomposition of the waste and the growth of the plants. You expect the yard waste to break down faster and produce taller plants. As a **control experiment,** you will grow one plant without any waste to judge the growth without compost. If the plant with yard waste compost grows taller than either of the other two plants, and the yard waste has decomposed more than the household waste, your hypothesis will be supported.

Level of Difficulty
Moderate, because of the time involved.

Materials Needed
- Three 2-gallon (7.5-liter) potting containers (terra cotta, ceramic, or plastic) with one or more holes in the bottom for drainage
- 3 pounds (1.3 kilograms) topsoil
- 3 to 5 pounds (1.3 to 2.3 kilograms) sand
- 3 to 5 pounds. (1.3 to 2.3 kilograms) organic waste (Use two types: household—table scraps, rotten vegetables, coffee grounds, etc.—and yard waste—leaves, twigs, grass clippings, weeds, etc.)
- 3 small identical living plants (annual flowers or vegetable plants), such as sunflowers, beans, or tomatoes
- 3 stakes for markers (Popsicle sticks will work)
- plastic or rubber gloves

How to Experiment Safely

Wear gloves when handling waste and mixing soil.

Approximate Budget

$5 (use topsoil from your yard if available).

Timetable

Two to four months.

Step-by-Step Instructions

1. Mix the topsoil and sand together to create the soil base.

2. Prepare the control experiment. Fill pot #1 with the soil base, leaving 2 inches (5 centimeters) at the top of the pot. Place one plant into the soil, covering all the roots. Water generously.

3. Prepare pot #2. Add to the soil base the household waste you collected (scraps, rotten vegetables, etc.). Mix the soil thoroughly. Place a plant into the soil, covering all the roots. Water generously.

LEFT: Step 2: How to fill pot #1.

RIGHT: Step 3: How to fill pot #2.

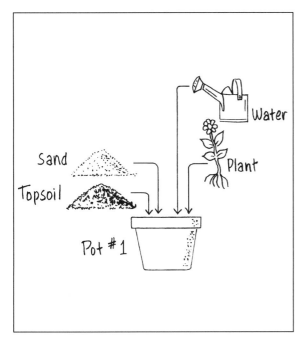

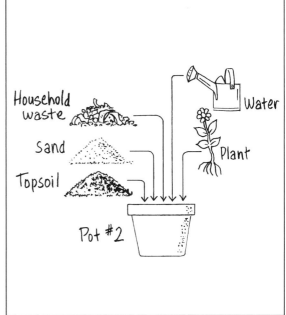

4. Prepare pot #3. Follow the directions for pot #2 but substitute the yard waste (grass clippings, leaves, etc.) instead of household waste.

5. Put markers in the pots identifying them as "control," "household," or "yard." Place the pots in a sunny location and monitor the growth of the plants. If possible, take photographs of them at the beginning of the experiment. Water the plants when the soil feels dry. Do not allow them to dry out completely.

6. Graph the weekly growth of the plants, recording the plant height, number of leaves, and root development, if visible.

7. After 2 to 4 months record the final heights and differences in the plant growth between each pot. Empty the pots and evaluate the amount of composting that occurred in the soil. Look for recognizable waste materials, record results.

Step 7: Sample plant height data sheet.

Week #	Plant height			Additional Observations (use other paper as needed)
	Control	Household	Yard	

Troubleshooter's Guide

Because of infinite variables, such as the different kinds of organic waste that you can use in this experiment, the result can vary greatly. For instance, if you use oak leaves, which are resistant to decay and highly acidic, your experiment's results may be different than expected. If one plant dies, the experiment should be restarted from the beginning. If you notice the leaves are being eaten, try to remove the pests, but first ask for help from an adult.

Summary of Results

During the experiment you will be recording the plant growth in the three pots. Ideally, the pot that is composting fastest will provide the most nutrients for its plant. It is essential to measure the height of each plant. You may also want to record which plant flowered first, how often it bloomed, and whether it produced fruit.

Change the Variables

Try varying the experiment by changing the variables. You can make two identical pots with the same soil, garbage, and plants. Give one pot half as much water as the other and compare the differences in growth. You can also experiment with the pH of the waste materials. Most leaves are acidic when composted and have a low pH. Try adding 1 cup (about 0.25 liter) of garden lime (calcium carbonate) to the soil to neutralize the acidic leaves.

Design Your Own Experiment

How to Select a Topic Relating to This Concept

To create your own experiment, consider your available resources. Decide what interests you. You may want to create a compost pile of household waste and create soil for a herb garden, or find ways to reduce your consumption of nonbiodegradable waste such as plastics. Although the choice is yours, you need a clear goal that will keep you motivated and interested.

Check the For More Information section and talk with your science teacher or school or community media specialist to start gathering information on composting questions that interest you.

Steps in the Scientific Method

To do an original experiment, you need to plan carefully and think things through. Otherwise, you might not be sure what question you are answering, what you are or should be measuring, or what your findings prove or disprove.

Here are the steps in designing an experiment:

- State the purpose of—and the underlying question behind—the experiment you propose to do.
- Recognize the variables involved, and select one that will help you answer the question at hand.
- State a testable hypothesis, an educated guess about the answer to your question.
- Decide how to change the variable you selected.
- Decide how to measure your results.

Recording Data and Summarizing the Results

It is important that your experiment's results are saved for other scientists to examine and compare. You should keep a journal and record notes and measurements in it. Your experiment can then be utilized by others to answer their questions about your topic.

Related Projects

When thinking about doing a project related to waste management, you need to limit your focus to one aspect of the field. For example, if you decide that recycling is your interest, choose what type of material you wish to work with. Since organic waste is smelly and metal and glass are dangerous, a good choice may be plastics. You can now begin to research ideas on how to recycle plastics. Recycling, composting, waste reduction, incineration, and conservation are all topics that can be explored and narrowed down to a concept that can lead to an interesting project.

For More Information

Franke, Irene, and David Brownstone. *The Green Encyclopedia*. New York: Prentice Hall, 1992. ❖ Good general reference book on environmental practices, including composting.

Leuzzi, Linda. *To the Young Environmentalist.* Stamford, CT: Franklin Watts, 1997. ❖ Interviews with respected environmentalists, including a biologist of a waste management facility.

Saunders, Tedd. *The Bottom Line of Green is Black: Strategies for Creating Profitable and Environmentally Sound Businesses.* New York: Harper and Row, 1992. ❖ Profiles of companies, such as Reader's Digest, that address landfill waste in their business practices.

chapter 9

Density and Buoyancy

What does it mean when it is said that one type of **matter** is more *dense* than another? What does **density** tell us about the nature and behavior of a substance? How does density affect the tendency of an object to float or sink in a liquid?

The density of matter is determined by the **mass** of a given **volume** of that matter. Any object at a given temperature and pressure will have a fixed volume, determined by the quantity of space it occupies and measured in cubic inches (cubic centimeters or milliliters). It also will have a fixed mass, determined by the quantity of matter contained in the substance. Mass is measured in pounds (kilograms). Density equals mass divided by volume.

The mass of different substances can vary greatly. The atoms that make up lead are tightly packed (at room temperature and pressure) and possess a large number of **subatomic particles**—protons, neutrons, and electrons. In contrast, the atoms that make up hydrogen gas are very loosely packed at the same temperature and pressure and possess a very small number of subatomic particles. More atoms with more subatomic particles in a given volume means higher density. Fewer atoms with fewer subatomic particles in a given volume means lower density.

Imagine a lifesize sculpture of a goldfish molded in solid clay. Now imagine an identical statue cast in solid lead. Both sculptures occupy the same volume, but the lead has a greater mass and is therefore denser. A third identical sculpture, this time carved from balsa wood, also occu-

Words to Know

Buoyancy:
The tendency of a fluid to exert a lifting force on a body immersed in it.

Density:
The mass of a substance divided by its volume.

Hypothesis:
An idea phrased in the form of a statement that can be tested by observation and/or experiment.

Immiscible:
Incapable of being mixed.

TOP: A ship floats in water because of the effects of density and buoyancy. (Photo Researchers Inc. Reproduced by permission.)

BOTTOM: Three statues of identical shape and size have different densities depending on their mass.

Words to Know

Mass:
Measure of the total amount of matter in an object. Also, an object's quantity of matter as shown by its gravitational pull on another object.

Matter:
Anything that has mass and takes up space.

Relative density:
The density of one material compared to another.

Specific gravity:
The density of a material compared to water.

Subatomic:
Smaller than an atom. It usually refers to particles that make up an atom, such as protons, neutrons, and electrons.

Variable:
Something that can affect the results of an experiment.

Volume:
The amount of space occupied by a three-dimensional object.

LEAD   CLAY   WOOD

pies the same volume but contains less mass than either the clay or the lead. Balsa wood is less dense than both clay and lead.

Density is measured on a relative scale

Notice that in comparing the densities of lead, clay, and balsa wood, we have not used any units of measurement. We simply stated that balsa wood is less dense and lead is more dense compared to clay. This is called **relative density**.

experiment
CENTRAL

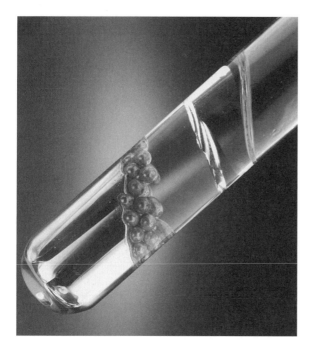

To measure density, scientists often use a relative scale. Water is assigned a value of 1.0, and other materials are assigned numerical values greater or less than 1.0 based on their density relative to water. For example, lead has a relative density of 11.3 and balsa wood has a relative density of 0.2. Relative density compared to water is also called **specific gravity.**

Relative density can be observed

The relative density of certain materials is easy to determine by observing the behavior of the materials when gravity acts upon them in a liquid. Substances of greater density will sink in liquids of lesser density due to the greater gravitational pull on the mass they contain. Conversely, substances of lesser density will rise. Thus, the lead goldfish will sink through water, while the balsawood goldfish will float. What about the clay goldfish? To predict its behavior, we would need to know its relative density.

When two **immiscible** liquids, such as oil and vinegar, are poured into a container, the less-dense liquid will float on top of the more-dense liquid. If a third liquid whose density falls between the first and second is poured into the container, it will form a layer between the other two liquids. A solid dropped into the container will sink through

LEFT: Materials placed together in a container will float or sink according to their relative density. (Photo Researchers Inc. Reproduced by permission.)

RIGHT: Archimedes studied the relationship between density and buoyancy. (Photo Researchers Inc. Reproduced by permission.)

ⓦords to Know

Waterline:
The highest point to which water rises on the hull of a ship. The portion of the hull below the waterline is under water.

the liquids of lesser density than itself, but it will float on the layer of the liquid whose density is greater than the solid's density.

Look! It floats

The relationship between density and **buoyancy** was studied in the third century B.C. by Archimedes, a Greek philosopher and inventor. The Archimedes Principle states that the lifting effect of a liquid on an object is equal to the weight of the liquid displaced by the object. Thus, if the object contains less mass than the mass of the displaced liquid, the object will float.

The Archimedes Principle is what makes steel ships float. If the mass of the displaced water—that is, the mass of the volume of water pushed aside by the hollow hull of the ship below the **waterline**—is greater than the mass of the entire ship, then the ship will float, even though steel has a relative density greater than 1.

The behavior of various materials under the effect of gravity can be observed and used to estimate their relative densities. In the first experiment, you will use such observations to create a relative density scale of your own. The experiment should ultimately help you predict the behavior of various materials, like the clay goldfish, according to their assigned density values.

The second experiment will examine the effect of increased pressure on a buoyant object containing a gas to see how changing the volume can change the buoyancy.

Experiment 1
Density: Can a scale of relative density predict whether one material floats on another?

Purpose/Hypothesis

In this experiment, you will first create a relative density scale for eight materials. Then you will use that information to predict whether one material will float on the other when any two of the materials are placed together in a container.

To begin the experiment, use what you know about relative density to make an educated guess about whether one material will float on

experiment
CENTRAL

What Are the Variables?

Variables are anything that could affect the results of an experiment. Here are the main variables in this experiment:

- the type and purity of the materials
- the method by which the materials are added to the container
- the order in which materials are added to the container
- the temperature at which the materials are kept
- the pressure at which the materials are kept

In other words, the variables in this experiment are everything that might affect the ability of one material to float on another. If you change more than one variable, you will not be able to tell which variable had the most effect.

the other. This educated guess, or prediction, is your **hypothesis**. A hypothesis should explain these things:

- the topic of the experiment
- the **variable** you will change
- the variable you will measure
- what you expect to happen

A hypothesis should be brief, specific, and measurable. It must be something you can test through observation. Your experiment will prove or disprove whether your hypothesis is correct. Here is one possible hypothesis for this experiment: "A relative density scale based on the behavior of eight materials in one container will accurately predict that a material with a lower relative density will float on one with a higher relative density when the two are placed in another container."

In this case, the variables you will change are the two materials, and the variable you will measure is which material floats on the other. If the material with the lower relative density floats on the one with a higher relative density, you will know your hypothesis is correct.

Level of Difficulty

Moderate.

Materials Needed

- 3 clear, narrow, glass jars with wide mouths (such as beakers or pickle jars)
- 1 probe (a knitting needle or drink stirrer will do)
- 9 disposable plastic knives
- corn oil
- motor oil (10W-30)
- maple syrup
- water, colored blue with food coloring
- lemon juice
- one 0.5-inch-diameter (1.2 centimeter) ball of clay
- one 0.5-inch-diameter (1.2 centimeter) ball of candle wax
- 1 small cork

Approximate Budget

Less than $10. (Most, if not all, materials may be found in the average household.)

Timetable

To be performed properly, allowing time for materials to settle and for careful observing and note taking, this experiment should take 45 to 60 minutes.

Step-by-Step Instructions

1. Divide your materials into liquids and solids. Examine the liquids first and try to predict which are the most dense. Pour the five liq-

 ## How to Experiment Safely

Before substituting other substances for those on the materials list, check with your science teacher to make sure you are not combining chemicals that will create a hazard, such as toxic fumes. Some combinations of household substances mix together easily or are the same color and therefore are not useful for this experiment. Throw away the knives and glass jars after finishing the experiment because they may be contaminated with motor oil.

experiment
CENTRAL

uids into one container, beginning with the one you predict to be the most dense. Pour each liquid slowly, using a plastic knife as a guide, as illustrated. Liquids that normally do not mix may accidentally mix if they are shaken or stirred. Use a new knife for each liquid.

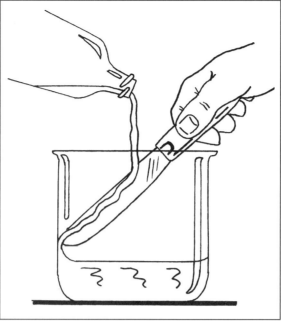

Step 1: Pouring liquid using a knife as a guide.

2. After all the liquids have been added to the container, wait for 1 minute to allow them to settle. Make a note of the order in which the liquids have settled, but do not assign relative density values yet. You have not yet added the solid materials, and the behavior of the solids may surprise you!

3. One by one, gently add the three solids to the container. Allow more time for them to settle. If a solid becomes coated with a liquid, its behavior may change temporarily. For example, a solid may float higher than normal if it is coated with vegetable oil. If you suspect that a solid is not behaving normally, gently poke it with the probe.

4. After you are confident that all the materials have settled to their natural levels, begin assigning relative density values. Start by identifying the layer of blue water and label that "1.0" on your relative density scale (see illustration). Then identify each material above and below the water, record it on your scale, and assign a relative density value for each. Your numerical values do not need to be exact as long as their relative values show which material is denser. For example, you could assign 0.9 to the material just above the water and 0.8 to the material just above that. Likewise you could assign 1.1 to the material just below the water, and so on.

Relative Density Scale

Material or Substance	Relative Density Value
cork	-1
water	0 (zero)
clay	1

5. Select two different materials and carefully pour or place them in the second glass jar, using a new plastic knife for each liquid. (Do not pair a solid with another solid.) Record the order in which you add each material. Observe the behavior of the materials in the jar. Did your relative density scale accurately predict what would happen? If so, your hypothesis has proven correct so far.

6. Determine whether the behavior of the materials used in the previous step changes when the order of putting the materials into the jar is changed. For example, if you previously added motor oil to a jar already holding water, now reverse the order, pouring the oil in first. Use the third jar and clean knives for this test.

Summary of Results

Examine your results and determine whether your hypothesis is correct. Did the observed behavior of the eight materials combined make it possible to create a useful relative scale? If any of the behaviors disagree with the scale's prediction, try to find a possible explanation for

experiment
CENTRAL

Troubleshooter's Guide

Here are some problems that may arise during the experiment, some possible causes, and ways to remedy the problems:

Problem: Two liquids appear to mix.

Possible causes:

1. Agitation when pouring the liquid into the container may cause temporary mixing. Wait for the mixture to settle out.

2. Two of your substances are too similar in appearance, such as vegetable oil and motor oil. Replace one substance with something that is similar but provides more contrast. For example, you could use canola oil in place of vegetable oil.

Problem: The behavior of a solid in liquids is erratic: sometimes it floats, sometimes it sinks.

Possible cause: Surface tension can sometimes cause an object of greater density to float on top of a liquid of lesser density. To counteract this tension, poke the solid with the probe.

this difference. Did you misread the layers in the first step of your experiment? Go back and double check. Write a summary of your findings.

Change the Variables

You can vary this experiment in several ways. Try different liquids and solids. Compare the densities of two solids, such as clay and a piece of pencil eraser. Then create and test a combination of solids by wrapping the eraser inside a layer of clay. Be sure to check with your teacher before trying new materials to make sure they are safe when mixed!

You can also see if you get the same results when the liquids in your experiment are chilled. (Do not heat your materials.) Freeze a liquid material and see if its relative density is the same whether in liquid or solid form.

experiment
CENTRAL

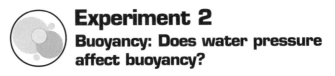

Experiment 2
Buoyancy: Does water pressure affect buoyancy?

Purpose/Hypothesis

In this experiment, you will observe the effect of increased water pressure on two buoyant objects floating in a closed bottle of water. The first is a flexible drinking straw filled with air and open at one end. The second is a flexible drinking straw filled with air and sealed at both ends. Because the first straw is open at one end, an increase in pressure allows water to easily force its way into the straw. This decreases the volume of water the straw displaces and it will eventually sink. Because the second straw is sealed at both ends, the water cannot force its way inside and must actually collapse the straw to decrease the displaced volume.

To begin the experiment, use what you know about buoyancy to make an educated guess about how the straws will behave when the pressure is increased. This educated guess, or prediction, is your **hypothesis**. A hypothesis should explain these things:

• the topic of the experiment
• the **variable** you will change
• the variable you will measure
• what you expect to happen

What Are the Variables?

Variables are anything that might affect the results of an experiment. Here are the main variables in this experiment:

• the rigidity or flexibility of the walls of the object

• the gas present inside the objects

• the liquid in which the objects are placed

• the pressure applied to the objects

In other words, the variables in this experiment are everything that might affect the buoyancy of the objects. If you change more than one variable, you will not be able to tell which variable had the most effect on the buoyancy.

experiment
CENTRAL

A hypothesis should be brief, specific, and measurable. It must be something you can test through observation. Your experiment will prove or disprove whether your hypothesis is correct. Here is one possible hypothesis for this experiment: "A flexible drinking straw, filled with air and sealed at one end, will lose its buoyancy and sink at a lower pressure than one sealed at both ends."

In this case, the variable you will change is the amount of pressure applied, and the variable you will measure is whether the straws sink. If the straw sealed at one end sinks at a lower pressure than the one sealed at both ends, you will know your hypothesis is correct.

Level of Difficulty
Moderate.

Materials Needed
- one 1-liter transparent plastic bottle filled with water (the bottle must have flexible sides and a cap that seals)
- 2 transparent drinking straws
- modeling clay
- 1 tall drinking glass
- water

Approximate Budget
Less than $5. (Most, if not all, materials may be found in the average household.)

Timetable
Approximately 10 to 20 minutes.

Step-by-Step Instructions
1. Cut a 4-inch (10-centimeter) length of straw and seal one end with a lump of clay. This will be the top end of the straw. Attach a ring of clay to the straw near the open bottom end to serve as ballast to

How to Experiment Safely
Make sure the bottle's cap is secured tightly before applying pressure.

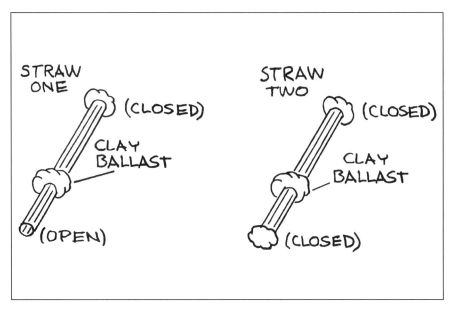

Steps 1 and 2: Set-up of straw 1 and straw 2.

keep it upright in the water as illustrated. Fill the drinking glass with water and test the buoyancy of the first straw. Add or remove clay from the ballast until the straw floats upright in a stable manner.

2. Repeat this process with the second straw, but seal this one with clay at both ends. Check the seals by submerging the top of the straw in the drinking cup. Look for bubbles coming from the top seal. Then invert the straw and check the bottom seal.

3. Fill the bottle with water to within 1 to 2 inches (2.5 to 5.0 centimeters) of the neck. Carefully lower the two straws into the bottle with the bottom end of the straws down. Close the bottle and make sure it is sealed tightly.

4. Position the bottle on a table or counter so that one person can squeeze the bottle while another takes measurements with the ruler of the change in the bottle's width where it is squeezed. This measurement will serve as a rough gauge of the pressure applied to the water and objects inside the bottle.

5. Measure and record the approximate diameter of the bottle. Gently squeeze the bottle until its width has decreased by 0.5 inch (1.25 centimeter). Record any change that occurs in the straws (sinking, taking on water, deforming) in the appropriate column on your data chart. Repeat this process for each 0.5-inch

narrow diameter of bottle (rough est. of pressure)	straw one (open end)	straw two (closed)
11 cm	floating at surface	floating at surface
10 cm	still floating at surface, water climbing 2 cm up into straw	still floating at surface, slight "crushing" of straw
9 cm	sinks to bottom, water 4 cm up into straw	still floating, straw "crushed" to 1/2 normal width

Note: These figures may vary depending on the size of the bottle used and the quantity of water.

Step 5: Sample recording chart.

(1.25-centimeter) change in the bottle's width. As increasing pressure is applied, the straw with the open end should sink.

6. Continue squeezing until the second straw sinks or until no more pressure can safely be applied to the bottle.

7. When pressure is released, the straw or straws should regain their buoyancy and return to the surface. Repeat the experiment, this time noting any changes you observe in the two straws as pressure is applied to the bottle. Watch for water rising in the unsealed straw. This is similar to a submarine flooding its ballast tanks to decrease its buoyancy and dive under water. Watch for deformation of the second straw, which should flatten as the pressure is increased.

8. Examine your results and determine whether your hypothesis is true. Repeat the experiment to double check your results. Write a summary of your findings.

Troubleshooter's Guide

Here are some problems that may arise during the experiment, some possible causes, and ways to solve the problems:

Problem: Neither straw sinks, even when maximum pressure is applied.

Possible causes:

1. The bottle may not be properly sealed. Check the seal. If necessary, place a small amount of clay on the threads of the bottle top to help keep a seal.

2. There is too much air in the bottle. Add water.

Problem: The first straw sinks, but the second does not.

Possible causes:

1. You are not applying enough pressure. Try having two people press on the bottle (carefully!) from either side.

2. The straws are too rigid. Use straws of less rigid plastic.

3. Your hypothesis is incorrect.

Problem: Once the straw or straws have sunk, they do not return to the surface when pressure is released.

Possible cause: The straw or straws are leaking. Check the clay seals.

Problem: The straw or straws are unstable and tend to flip over.

Possible cause: The ballast weight is not heavy enough or is not placed properly. Increase the weight or move the ballast weight farther down the straw.

Summary of Results

Record your data on a chart. This chart should be as clear as possible. It will contain the information that will show whether your hypothesis is correct.

Change the Variables

You can vary this experiment. Here are some possibilities. Try different numbers and lengths of straw. Compare the behavior of short straws and long straws. See if you get the same results with different liquids. Try salt water and carbonated water.

 # Design Your Own Experiment

How to Select a Topic Relating to This Concept

Demonstrations of the properties of density and buoyancy exist in our environment in numerous forms. Everyday sights such as helium balloons floating away or a thin slick of oil on a roadside puddle show the principles we have investigated in our experiments. Think of ways to vary the conditions you observe that will answer questions you have about buoyancy.

Check the For More Information section and talk with your science teacher or school or community media specialist to start gathering information on density and buoyancy questions that may interest you. As you consider possible experiments, be sure to discuss them with your science teacher or another knowledgeable adult before trying them. Some materials or procedures might be dangerous.

Steps in the Scientific Method

To do an original experiment, you need to plan carefully and think things through. Otherwise, you might not be sure which question you are answering, what you are or should be measuring, or what your findings prove or disprove.

Here are the steps in designing an experiment:

- State the purpose of—and the underlying question behind—the experiment you propose to do.
- Recognize the variables involved and select one that will help you answer the question at hand.
- State your hypothesis, an educated guess about the answer to your question.
- Decide how to change the variable you have selected.
- Decide how to measure your results.

Recording Data and Summarizing Results

In the experiments included here and in any experiments you develop, you can look for better ways to display your data in more accurate and interesting ways. For example, in the buoyancy experiment, try to find a better way to measure the pressure inside the bottle. Could a pressure gauge be built into the bottle's cap without altering the results?

Remember that those who view your results may not have seen the experiment performed, so you must present the information you have gathered in as clear a way as possible. Including photographs or illustrations of the steps in the experiment is a good way to show a viewer how you got from your hypothesis to your conclusion.

Related Projects

Although experiments in density and buoyancy can be challenging and fun, simple demonstrations of the principles involved can also be highly informative and often can reveal surprising facts. Many aspects of density and buoyancy, such as the effect of salinity, could yield interesting experimental results.

For More Information

Gillett, Kate, ed. *The Knowledge Factory.* Brookfield, CT: Copper Beech Books, 1996. ❖ Provides some fun and enlightening observations on questions relevant to this topic, along with good ideas for projects and demonstrations.

Ray, C. Claibourne. *The New York Times Book of Science Questions and Answers.* New York: Doubleday, 1997. ❖ Addresses both everyday observations and advanced scientific concepts on a wide variety of subjects.

Wolke, Robert L. *What Einstein Didn't Know: Scientific Answers to Everyday Questions.* Secaucus, NJ: Birch Lane Press, 1997. ❖ Contains a number of interesting entries on the nature of water.

Dissolved Oxygen

What turns a body of water into a "dead zone" where nothing can live? One condition that can wipe out most living things in a stream, river, or lake is a low level of **dissolved oxygen.** The term dissolved oxygen refers to molecules of oxygen that have been dissolved in water. Some of these molecules enter the water from the surrounding air, especially if the water tumbles over falls and rapids. Other dissolved oxygen in the water is a "by-product" of **photosynthesis.** During photosynthesis, green plants, including those that live in the water, use the energy in sunlight to combine carbon dioxide and water to produce carbohydrates and oxygen. The oxygen is expelled by the plant and enters the water.

The level of dissolved oxygen in water can reach as high as 8 or 9 parts per million. The United States Environmental Protection Agency (EPA) considers water to be healthy if it contains at least 5 parts per million of dissolved oxygen. When the level falls below 4 parts per million, the water quality is considered to be poor. At 2 parts per million, fish become stressed and grow more slowly, and some die.

What affects the level of dissolved oxygen in water?

The level of dissolved oxygen in a body of water can vary from hour to hour. The level falls as fish remove oxygen molecules from the water with their gills. The more fish in the water, the more dissolved oxygen they remove. Fish are cold-blooded, so their body systems work more slowly in cold water and speed up in warm water. The warmer the water, the more oxygen their body systems require. Plants in the water, including the tiny floating **phytoplankton,** also use small amounts of the dissolved oxygen for **respiration.**

(W)ords to Know

Biochemical oxygen demand (BOD_5):
The amount of oxygen microorganisms use over a five-day period in 68°F (20°C) water to decay organic matter.

By-product:
A secondary substance produced as the result of a physical or chemical process, in addition to the main product.

Control experiment:
A set-up that is identical to the experiment but is not affected by the variable that will be changed during the experiment.

experiment
CENTRAL

Photosynthesis requires sunlight, so plants do not produce oxygen at night. During these dark hours, plants actually use more oxygen for respiration than they produce. That's why the level of dissolved oxygen in a body of water is lowest just before dawn, just before the Sun rises and photosynthesis begins again. If you visit a pond or river at dawn, you might see birds picking fish out of the water. The fish are easy to catch then because they are at the surface, gulping for oxygen because the water does not provide enough for them.

Other factors also influence the level of dissolved oxygen, including the water's temperature, its **salinity,** and its **elevation** above sea level. As the water temperature decreases, the amount of dissolved oxygen increases, because gases, including oxygen, dissolve more easily in cooler water. As the level of salinity increases, the amount of dissolved oxygen decreases. Finally, bodies of water at higher elevations, such as mountain lakes, contain less dissolved oxygen than bodies of water at lower elevations. This makes sense when you remember that much of the dissolved oxygen comes from the air. The amount of oxygen in the air decreases the higher you climb on a mountain. If the air has less dissolved oxygen, the water will, too.

During hot, dry summer months, the water level in streams tends to be low, and the water often becomes stagnant. The heat and the lack of movement combine to lower dissolved oxygen levels. On the other hand, during the early spring, melting snow and cool rain keep the water temperatures low, increasing the dissolved oxygen levels. The rains lead to rushing, tumbling streams that gain more oxygen from the atmosphere. The rains also contribute the oxygen they absorbed from the atmosphere.

Another major effect on the level of dissolved oxygen in a body of water is the amount of pollutants in the water. Many pollutants, including the fertilizers that run off farm fields and home lawns, contain nutrients that help plants grow, including plants in the water. This may seem like a benefit of pollution. However, after the plants use up the nutrients in the water, they die and start to decay. The bacteria involved in the decay process use the dissolved oxygen in the water, reducing the amount of oxygen available to the fish. This process is called **eutrophication.** As eutrophication continues to use up the dissolved oxygen, the water can turn into a dead zone.

Scientists have measured the **biochemical oxygen demand,** the amount of oxygen required by bacteria to decay waste material. BOD_5

OPPOSITE PAGE:
As more water surface is exposed to the air, more oxygen molecules enter the water. (Photo Researchers Inc. Reproduced by permission.)

(**W**)**ords to Know**

Dissolved oxygen:
Oxygen molecules that have dissolved in water.

Elevation:
Height above sea level.

Eutrophication:
Natural process by which a lake or other body of water becomes enriched in dissolved nutrients, spurring aquatic plant growth.

Hypothesis:
An idea in the form of a statement that can be tested by observation and/or experiment.

Photosynthesis:
Chemical process by which plants containing chlorophyll use sunlight to manufacture their own food by converting carbon dioxide and water into carbohydrates, releasing oxygen as a by-product.

Phytoplankton:
Microscopic aquatic plants that live suspended in the water.

experiment
CENTRAL

means the amount of oxygen that microorganisms use to decay organic matter over a five-day period in 68°F (20°C) water. The more waste in the water, the more decay that occurs, and the higher the BOD_5—the need for dissolved oxygen. For example, wastewater that has been treated has a BOD_5 of less than 30 parts per million. However, waste from a meat packing plant has a BOD_5 of 5,000 parts per million. If this meat packing waste were released into a body of water, the dissolved oxygen level in that water would drop dramatically within a few days.

How does a low level of dissolved oxygen affect the ecosystem in the water?

If the level of dissolved oxygen drops for any length of time, fish that need large amounts of oxygen, such as trout and bass, go elsewhere if they can. Carp, catfish, worms, and fly larvae (the immature, worm-like stage in a fly's life cycle) can handle low oxygen levels, so they thrive. The ecosystem begins to include more organisms that can live

A pond overrun with algae is usually not a healthy place. (Photo Researchers Inc. Reproduced by permission.)

OPPOSITE PAGE:
At high altitudes, cold temperatures raise the level of dissolved oxygen, but the higher elevation lowers it. The level of dissolved oxygen in any body of water is a complex, changing condition. (Photo Researchers Inc. Reproduced by permission.)

with little or no oxygen. If the level of dissolved oxygen continues to drop, even the carp and catfish end up gasping for oxygen. The water is on its way to becoming a dead zone.

In the following two experiments, you will use a kit to measure the level of dissolved oxygen in water under several conditions. In one experiment, you will determine how the level changes as the amount of decaying matter in the water changes. In the second experiment, you will measure how the breathing rate of goldfish changes as the amount of dissolved oxygen in the water changes. Both experiments will help you better understand the concept of—and the importance of—dissolved oxygen.

Experiment 1
Decay and Dissolved Oxygen: How does the amount of decaying matter affect the level of dissolved oxygen in water?

Purpose/Hypothesis
In this experiment, you will allow different amounts of food to decay in water and measure any changes that occur in the level of dissolved oxygen.

To begin the experiment, use what you have learned about dissolved oxygen to make a guess about what will happen when the food starts to decay in the water. Will the level of dissolved oxygen in the water decrease or increase? Will the amount of change depend on the amount of decaying food? This educated guess, or prediction, is your **hypothesis.** A hypothesis should explain these things:

- the topic of the experiment
- the **variable** you will change
- the variable you will measure
- what you expect to happen

A hypothesis should be brief, specific, and measurable. It must be something you can test through observation. Your experiment will prove or disprove whether your hypothesis is correct. Here is one possible hypothesis for this experiment: "The more decaying matter in the water, the lower the level of dissolved oxygen."

What Are the Variables?

Variables are anything that might affect the results of an experiment. Here are the main variables in this experiment:

- the beginning levels of dissolved oxygen in each container

- the amount of decaying food in each container of water

- how much the food is decayed

- the temperature of the water in all containers

- the amount of any mixing, pouring, or splashing of the water in the containers (which would raise the dissolved oxygen level)

- the length of time the containers are allowed to sit

In other words, the variables in this experiment are everything that might affect the level of dissolved oxygen. If you change more than one variable at a time, you will not be able to determine which variable affected the results.

In this case, the variable you will change is the presence and amount of decaying food, and the variable you will measure is the level of dissolved oxygen. As a **control experiment,** you will set up one container of water with no decaying food in it. That way, you can determine whether the level of dissolved oxygen changes even with no decaying food in the water. If the level of dissolved oxygen decreases with an increase in decaying food and does not change in the control container, your hypothesis is correct.

Level of Difficulty

Easy/moderate.

Materials Needed

- 3 clear 0.5-gallon (1.9-liter) containers
- about 3 ounces (85 grams) of rotting fruit, such as brown apple slices or an overripe banana

- scale capable of weighing 2 ounces (57 grams)
- dissolved oxygen test kit (kits are available from biological supply houses; one popular brand is LaMotte; see the For More Information section for sources)
- 1.5 gallons (5.6 liters) water (try to obtain water that has not been treated, such as well, stream, or pond water; many water treatment plants try to reduce the level of dissolved oxygen in their water because high levels speed up corrosion in water pipes)
- wax paper
- goggles
- rubber gloves

Approximate Budget
$15-$20 for the test kit and $5 for a small food scale; other materials should be available in the average household.

Timetable
15 minutes to set up; 1 week to observe.

Step-by-Step Instructions

1. Label the containers "1 oz.," "2 oz.," and "Control."

2. Mix your water supply thoroughly; stir the water vigorously for 5 minutes or more if you used tap water, which tends to have a low dissolved oxygen level.

3. Nearly fill the three containers with the water.

4. Follow the directions on the water testing kit to measure the beginning level of dissolved oxygen in each container. Record the levels in a chart similar to the one illustrated. (The water in all three containers should have the same dissolved oxygen level at this point.)

 How to Experiment Safely

Wear goggles and gloves to protect your eyes and skin while you test the water because you will be using chemicals that can be dangerous. You are strongly urged to have an adult help you complete the tests.

Dissolved Oxygen Levels

	1-oz container	2-oz container	control
Beginning of Experiment			
Day 2			
Day 3			
Day 4			
Day 5			

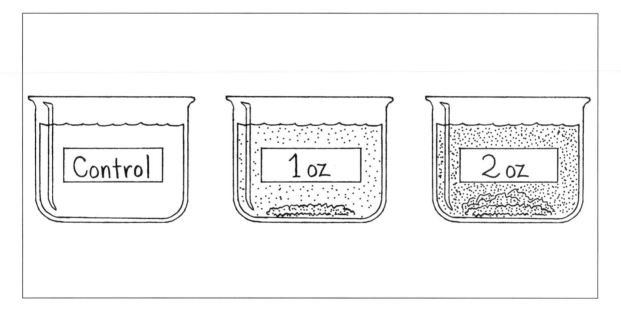

5. Put wax paper on the scale and measure 1 ounce (28 grams) of rotting fruit; dump the fruit into the container marked "1 oz."

6. Measure 2 ounces (57 grams) of the same rotting fruit and dump it into the container marked "2 oz." Put no fruit in the control container.

7. Place all three containers in an area where the air temperature will remain at 70 to 72°F (21 to 22°C).

TOP: Step 4: Dissolved Oxygen Levels recording chart.

BOTTOM: Steps 5 and 6: Set-up of control, 1 oz., and 2 oz. containers.

8. Every day at the same time for the next four days, use the kit to test the dissolved oxygen level in each container. Record your findings on your chart. Also note the condition of the water. Are any of the containers becoming cloudy?

Summary of Results

Study the data from your observations and decide whether your hypothesis was correct. How did the dissolved oxygen levels change in the three containers? Which container had the highest level at the end of the experiment? The lowest level? Did the level change in the control container? If so, why do you think this happened? Write a para-

Troubleshooter's Guide

Below are some problems that may arise during this experiment, some possible causes, and ways to remedy the problems.

Problem: The level of dissolved oxygen was really low in all three containers in the beginning.

Possible cause: Your water came from a source with little dissolved oxygen. Try the experiment again, but increase the beginning level of dissolved oxygen by running a tube from an aquarium pump into the water. Send bubbles of air through the water for at least 8 to 12 hours. Treat all the water so the beginning levels will be identical in all containers.

Problem: The level of dissolved oxygen dropped in all containers, including the control.

Possible cause: The water already had some decaying matter in it, especially if it was pond water. Focus on the differences in the levels of dissolved oxygen for all three containers.

Problem: The level of dissolved oxygen rose in the control container.

Possible cause: The room temperature cooled enough so that oxygen from the air entered the water. Make sure the temperature around all three containers stays at 70 to 72°F (21 to 22°C).

graph summarizing your findings and explaining whether they support your hypothesis.

Change the Variables

You can vary this experiment in several ways. For example, you might try a different kind of decaying matter, such as another kind of fruit, raw meat, moldy bread, or rotting leaves. You could also increase or decrease the air temperature around all three containers to see how that affects the rate of decay and the levels of dissolved oxygen. At the end of the experiment, use aquarium pumps and tubing to bubble the same amount of air into all three containers to try to raise the level of dissolved oxygen. To change the salinity of the water, you could add different amounts of salt to two containers instead of decaying food and measure any changes in the levels of dissolved oxygen.

Experiment 2

Goldfish Breath: How does a decrease in the dissolved oxygen level affect the breathing rate of goldfish?

Purpose/Hypothesis

In this experiment, you will observe the breathing rate of goldfish as they swim in water with different levels of dissolved oxygen. [NOTE: It is recommended that you perform this experiment only if you already have access to an aquarium with four to six goldfish and only with the permission of a responsible adult. This experiment will not harm the fish as long as you limit the duration of the experiment and return the fish to the main aquarium afterwards.]

To begin the experiment, use what you know about dissolved oxygen and its effect on fish to make an educated guess about how the fishes' breathing rate will change as the level of dissolved oxygen drops. This educated guess, or prediction, is your **hypothesis**. A hypothesis should explain these things:

- the topic of the experiment
- the **variable** you will change
- the variable you will measure
- what you expect to happen

What Are the Variables?

Variables are anything that might affect the results of an experiment. Here are the main variables in this experiment:

- the health and size of all the goldfish
- the temperature and cleanliness of all the water
- the level of dissolved oxygen in the different containers of water

In other words, the variables in this experiment are everything that might affect the breathing rate of the fish. If you change more than one variable at a time, you will not be able to determine which change had more effect on your results.

A hypothesis should be brief, specific, and measurable. It must be something you can test through observation. Your experiment will prove or disprove whether your hypothesis is correct. Here is one possible hypothesis for this experiment: "As the dissolved oxygen level drops, the breathing rate of the goldfish will increase."

In this experiment the variable you will change is the level of dissolved oxygen, and the variable you will measure is the breathing rate of the goldfish. As a control experiment, you will observe the breathing rate of goldfish in an aquarium that has been set up for some time and in which the dissolved oxygen remains relatively constant. If the breathing rate of the control goldfish does not change, but the breathing rate of the other goldfish increases as the dissolved oxygen level drops, your hypothesis is correct.

Level of Difficulty

Easy/moderate.

Materials Needed

- one 10-gallon (38-liter) or larger aquarium that has been set up for a month or longer and uses an air pump to constantly bubble air through the water (the aquarium may also include live plants, which add more dissolved oxygen to the water; other fish living in

the aquarium will not affect the experiment, as long as they have been there for several weeks)
- one 1/2-gallon (1.9-liter) container
- 4 to 6 small goldfish
- dissolved oxygen test kit (see the For More Information section for sources)
- stopwatch
- fish net
- red and blue colored pencils
- goggles
- rubber gloves

Approximate Budget
$15 to $20 for the test kit. (Ideally, you will be able to use an aquarium that is already set up at school or at home.)

Timetable
15 minutes to set up the small container; 20 minutes to check the dissolved oxygen levels and breathing rates every 2 hours for 6 hours.

Step-by-Step Instructions
1. If you have to purchase additional goldfish to conduct the experiment, place them in the aquarium and allow 24 hours for them to get used to the water. During this period, if the aquarium has a heater, turn it off and allow the water to reach air temperature. Make sure the air pump continues to work.

2. Using water from the aquarium, fill the 1/2-gallon container.

 ## How to Experiment Safely
Treat the goldfish gently; avoid putting them into water that is warmer or cooler than they are used to. Limit the duration of the test to no more than 8 to 10 hours. Wear goggles and gloves to protect your eyes and skin while you test the water because you will be using chemicals that can be dangerous. You are strongly urged to have an adult help you complete the tests.

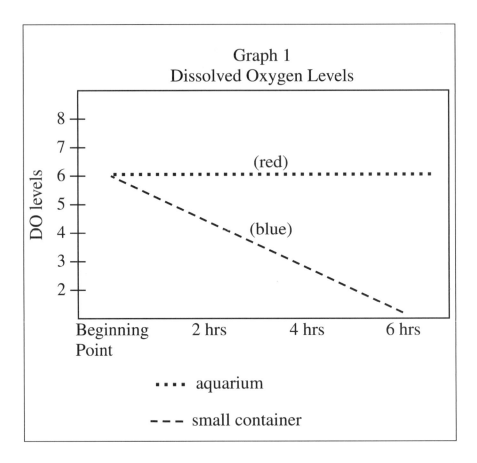

Graph 1
Dissolved Oxygen Levels

3. Use the kit to test the dissolved oxygen level in the aquarium and in the 1/2-gallon container. They should be the same at this point. On a graph similar to that illustrated, record the level from the aquarium in red and the level from the small container in blue.

4. Use the net to catch half of the goldfish (2 or 3); put them in the smaller container.

5. Use the stopwatch to measure how many times each goldfish breathes in 30 seconds. Each outward push of the gills is one breath. Average the breathing rates for the goldfish in the aquarium. Use the red pencil to record the average on a graph similar to that illustrated. Then average the breathing rates for the goldfish in the small container, and use the blue pencil to record that average on the graph.

6. Wait 2 hours and retest the dissolved oxygen levels in both containers. Then average the breathing rates of the fish in each container. Record your findings.

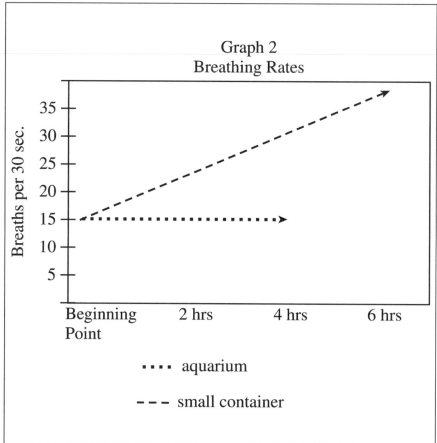

Step 4: Put 2 to 3 goldfish into smaller container.

Step 5: Sample graph of goldfish breathing rates.

7. Repeat Step 6 after 4 hours and after 6 hours.

8. At the end of the experiment, gently put the goldfish from the small container back into the aquarium. If you disconnected the aquarium heater, plug it back in.

Summary of Results

Study the dissolved oxygen levels on the first graph. What do you notice? Did the levels change in the aquarium? Did they change in the small container? If so, why?

Now compare the breathing rates of the two groups of fish, shown on the second graph. Notice whether the breathing rates changed as the levels of dissolved oxygen changed. How did the goldfish respond to any changes in the levels of dissolved oxygen? Was your hypothesis correct? Write a paragraph summarizing your findings and explaining whether they support your hypothesis.

Troubleshooter's Guide

Below are some problems that may arise during this experiment, some possible causes, and ways to remedy the problems.

Problem: The dissolved oxygen level in the small container remained the same.

Possible cause: The fish were too small to affect the level during this time period. Try the experiment again, using bigger or more fish, putting them in a smaller container of water, or extending the time period for the testing to 8 or 10 hours.

Problem: The breathing rate of the fish in the aquarium and the container dropped.

Possible cause: The water temperature might have fallen enough to slow the body processes of the goldfish. If possible, move the aquarium and small container to a warmer spot. Or leave the aquarium heater plugged in and put a heater in the small container to keep the water at the same temperature as the aquarium—a difficult feat to accomplish.

Change the Variables

You can vary this experiment in several ways. Measure and compare any change in the breathing rates of goldfish swimming in water with and without live plants. (Disconnect any air pump so the plants are the only source of added dissolved oxygen.) Or you can bubble air through the water in the small container and measure the breathing rate of the goldfish as the level of dissolved oxygen rises.

 # Design Your Own Experiment

How to Select a Topic Relating to this Concept

Measuring the amount of dissolved oxygen in a body of water is one of the best ways to determine the health of that water system and the environment around it. Consider the water sources near your home or school. Which ones might have high or low levels of dissolved oxygen? What might cause the high or low levels? What approaches might raise a low level? What other factors affect the health of a water system? (Examples include the pH level and the levels of ammonia, nitrates, and nitrites.)

Check the For More Information section and talk with your science teacher or school or community media specialist to gather information on dissolved oxygen questions that interest you. As you consider possible experiments, be sure to discuss them with a knowledgeable adult before trying them. Some of the materials or procedures may be harmful to yourself or to the environment.

Steps in the Scientific Method

To do an original experiment, you need to plan carefully and think things through. Otherwise, you might not be sure which question you are answering, what you are or should be measuring, or what your findings prove or disprove.

Here are the steps in designing an experiment:

- State the purpose of—and the underlying question behind—the experiment you propose to do.
- Recognize the variables involved, and select one that will help you answer the question at hand.

- State a testable hypothesis, an educated guess about the answer to your question.
- Decide how to change the variable you selected.
- Decide how to measure your results.

Recording Data and Summarizing Results

In your decaying food and goldfish experiments, your raw data might include charts, graphs, drawings, and photographs of the changes you observed. If you display your experiment, make clear the question you are trying to answer, the variable you changed, the variable you measured, the results, and your conclusions. Explain what materials you used, how long each step took, and other basic information.

Related Projects

You can undertake a variety of projects related to dissolved oxygen and water quality in general. For example, if you have access to salt water from the ocean, you might compare its level of dissolved oxygen with that of fresh water. Or compare the level at the surface of a pond with the level at the bottom. Or compare the level of dissolved oxygen in a body of water during cool weather with the level during a heat wave. Try to determine the factors that influence these levels and whether the levels indicate pollution that is potentially harmful to the health of the organisms living in the water and the people using and drinking it.

For More Information

Dashefsky, Steven. *Environmental Science: High-School Science Fair Experiments.* Blue Ridge Summit, PA: TAB Books, 1994. ❖ Focuses on more involved experiments relating to ecology and the environment.

Fitzgerald, Karen. *The Story of Oxygen.* Danbury, CT: Franklin Watts, 1996. ❖ Covers the history of oxygen, its chemistry, how it works in our bodies, and its importance in our lives.

Rybolt, Thomas, and Robert Mebane. *Environmental Experiments About Life.* Hillside, NJ: Enslow Publishers, 1993. ❖ Describes simple experiments relating to ecology.

Biological supply companies for dissolved oxygen test kits:

Carolina Biological Supply Company
2700 York Road
Burlington, NC 27215
1-800-334-5551

experiment
CENTRAL

Frey Scientific
100 Paragon Parkway
Mansfield, OH 44903
1-800-225-FREY

Ward's Natural Science Establishment, Inc.
5100 West Henrietta Road
PO Box 92912
Rochester, NY 14692
1-800-962-2660

LaMotte water test kits: http://www.lamotte.com/

chapter 11

Earthquakes

According to the ancient Greeks, **earthquakes** occurred when the god Atlas shifted the weight of the world from one shoulder to the other. Other cultures believed that earthquakes were a sign of punishment. We now know that earthquakes are the shaking or trembling of the earth caused by underground shock waves or vibrations. Believe it or not, over a million earthquakes take place each year. Sometimes the trembling and shaking is gentle and hardly noticeable. Other times the motion is much more violent, causing cracks in the surface of the earth.

There's a whole lot of shaking going on

Huge blocks of rocks called **plates** make up Earth's outer shell, or crust. These plates fit together like a cracked egg shell. The plates push and pull on each other constantly. Sometimes this pressure causes a **fault,** or a break in the rocks. Large pieces of these rocks, called **fault blocks,** can overlap. Pressure pushes on the rocks for centuries, finally causing them to rupture and snap in one big surge, resulting in a major earthquake.

Like a chain reaction, force from the movement of the rocks results in vibrations of the surrounding ground. These vibrations, or **seismic waves,** (pronounced SIZE-mic; relating to earthquakes) travel away from the break. Strong shaking from these waves lasts from 30 to 60 seconds and can cause buildings and highways to collapse.

Earthquakes can actually be beneficial. The constant shifting and upheaval of Earth's crust builds mountains and highlands. The planet would be flat without them.

ⓌWords to Know

Earthquake:
An unpredictable event in which masses of rock suddenly shift or rupture below Earth's surface, releasing enormous amounts of energy and sending out shockwaves that sometimes cause the ground to shake dramatically.

Epicenter:
The location where the seismic waves of an earthquake first appear on the surface, usually almost directly above the focus.

Fault:
A crack running through rock as the result of tectonic forces.

Violent earthquakes can cause cracks in the earth. (Photo Researchers Inc. Reproduced by permission.)

Words to Know

Fault blocks:
Pieces of rock from Earth's crust that press against each other and cause earthquakes when they suddenly shift or rupture from the pressure.

Focus:
The point within Earth where a sudden shift or rupture occurs.

Developing a theory

On November 1, 1755, the port of Lisbon, Portugal, was hit by a terrible earthquake. More than 60,000 people died. The day of the earthquake was a religious holiday, and many of those killed were crushed in churches. Because earthquakes were thought to be a punishment from God, it did not make sense that one would take place on a holy day. People also asked why innocent children would be punished? Soon after the earthquake, some people started to look for scientific reasons. The Marquez de Pombal, a Portugese nobleman, asked Lisbon's surviving priests to fill out questionnaires documenting information about the earthquake. The questionnaires included questions about the time and the direction of the earthquake shock.

In 1760, John Michell, an English physicist, came up with an interesting theory. He reasoned that if you could record the underground shock waves and the points at which the waves stopped, you could determine the point of origin, or **epicenter**, of an earthquake.

Epicenters existed deep in the rocks beneath the sea, he said. His theories, which were fairly accurate, were the start of **seismology,** the science of earthquakes and their origins.

Measuring an earthquake

In the first century, Chang Heng—a Chinese astronomer, mathematician, and writer—invented the earliest earthquake recorder. This device measured the occurrence and direction of an earthquake's motion. Italian physicist Luigi Palmieri has been credited with inventing the first **seismograph** in 1855. Seismographs detect and record earthquake waves. To pinpoint how dangerous an earthquake was, American seismologist Charles F. Richter (1900–1985) began measuring the peaks and valleys of these waves in the 1930s. He came up with a mathematical formula, known as the Richter (pronounced RIK-ter) Scale, which measures earthquake magnitude on a scale from 1 to 10. The Richter Scale also measures how much energy is released in an earthquake.

In the famous Lisbon, Portugal, earthquake of 1755, residents were killed by toppling buildings, fires, and high waves. (Corbis/Bettmann. Reproduced by permission.)

Words to Know

Plates:
Huge blocks of rocks that make up Earth's outer shell and fit together like a cracked egg.

Seismic waves:
Vibrations in rock and soil that transfer the force of an earthquake from the focus into the surrounding area.

Seismograph:
A device that detects and records vibrations of the ground.

Seismology:
The study and measurement of earthquakes.

Tectonic:
Relating to the forces and structures of the outer shell of Earth.

Tsunami:
A large wave of water caused by an underwater earthquake.

Variable:
Something that can affect the results of an experiment.

Increasing one whole number on the Richter Scale, from 5.0 to 6.0 for example, represents an increase of ten times the magnitude and about sixty times the energy.

Earth is a dynamic and changing planet. Conducting experiments will help you understand how earthquakes are part of the changes that are taking place.

Experiment 1
Detecting an Earthquake: How can movement of Earth's crust be measured?

Purpose/Hypothesis

In this experiment, you will construct a simple seismograph and simulate the forces that cause an earthquake. Your seismograph is a simple model, but you will see if it can detect vibrational activity in your house or building.

You probably have an educated guess about the outcome of this experiment based on what you already know about earthquakes. This educated guess, or prediction, is your **hypothesis.** A hypothesis should explain these things:

- the topic of the experiment
- the **variable** you will change
- the variable you will measure
- what you expect to happen

A hypothesis should be brief, specific, and measurable. It must be something you can test through observation. Your experiment will

experiment
CENTRAL

What Are the Variables?

Variables are anything that might affect the results of an experiment. Here are the main variables in this experiment:

- the amount of simulated earthquake disturbance
- the distance of the disturbance from the seismograph
- the surface on which you place your seismograph

In other words, the variables in this experiment are everything that might affect the amount of disturbance recorded on your seismograph. If you change more than one variable, you will not be able to tell which variable had the most effect on the seismograph recordings.

prove or disprove whether your hypothesis is correct. Here is one possible hypothesis for this experiment: "By simulating an earthquake with various types of disturbances, you will detect and record various types of vibrational activity on your seismograph."

In this case, the variable you will change is the amount of simulated earthquake disturbance, and the variable you will measure is the amount of displacement recorded on your seismograph. If a greater simulated disturbance results in a greater displacement on your seismograph, you will know your hypothesis is correct.

Level of Difficulty

Moderate. (The design of your seismograph is easy, but you may need someone to hold some pieces while you attach them. Also, you will need help from friends in creating vibrations.)

Materials Needed

- 1 or 2 helpers
- cardboard box about 12 inches x 12 inches (30 centimeters x 30 centimeters) with an opening on top
- scissors
- ruler

experiment
CENTRAL

- adding machine tape
- string
- pencil (or dowel)
- 5-ounce (about 148-milliliter) paper cup
- masking tape
- black marking pen
- small rocks or marbles
- modeling clay

Approximate Budget

$3.

Timetable

One hour.

Step-by-Step Instructions

1. Turn the box on its side so the opening is facing outward.

LEFT: Steps 2 to 4: Initial set-up of seismograph box.

2. Cut a 2-inch (5-centimeter) circle in the center of the top side of the box.

RIGHT: Steps 7 to 12: Completion of seismograph box.

3. Cut two 1/2-inch x 4-inch (1.25-centimeter x 10-centimeter) slots in the box. The first slot should be in the center of the bottom,

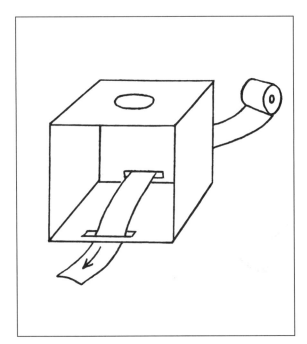

near the front opening. The second slot should be in the back center near the bottom. See the illustration.

4. Thread the adding machine tape through the slots, so the leading edge comes out the front slot.

5. Cut two 24-inch (61-centimeter) lengths of string.

6. Use the point of a pencil to poke two holes below the rim of the cup opposite each other.

7. Tie one string onto each hole in the cup.

8. Bring the free ends of the string through the 2-inch (5-centimeter) circle in the top side of the box.

9. Tape or tie the ends of the string to the pencil and lay the pencil across the hole.

10. Push the marking pen through the bottom of the cup, tip down.

11. Fill the cup with the rocks or marbles.

12. Adjust the height of the cup/pen/rock device so the marker tip just touches the adding machine tape. (You can adjust the string on the pencil, then fix the pencil in place using the modeling clay and masking tape.)

13. Test the device by pulling the adding machine tape forward with one hand and shaking the box gently with the other and observe the markings left on the paper.

14. Perform a seismic test indoors. Place your seismograph on the floor in the middle of the room. Have several of your friends walk, skip, jog, and run around in the room in a circle, always keeping the same distance away from the seismograph. While they are moving

experiment
CENTRAL

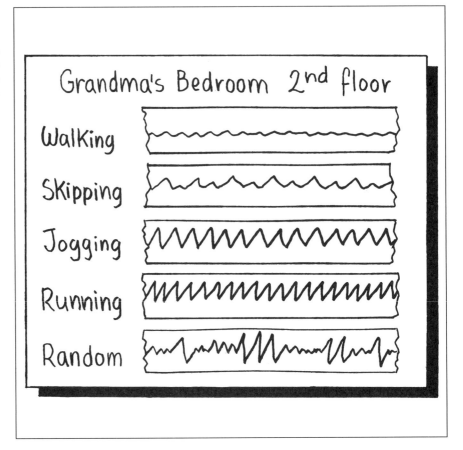

Grandma's Bedroom 2nd floor

Walking

Skipping

Jogging

Running

Random

about, record the seismic waves, or seismic activity, by slowly pulling the adding machine tape through the instrument (see illustration).

15. Label the tape with the location and activities.

Summary of Results

Compare your tapes. Do they show greater movement when the activity was more vigorous? In other words, does your seismograph accurately detect and record seismic activity?

Change the Variables

You can change one of the variables and repeat this experiment. For example, you can have your friends move closer or farther away from the seismograph to determine how the recordings vary. You can also place the seismograph on a shaky table, like an old card table, to see if this amplifies the disturbances.

Troubleshooter's Guide

Experiments do not always work out as planned. Below are some problems that may arise during this experiment, some possible causes, and ways to remedy the problems.

Problem: Nothing is being recorded on the adding machine tape.

Possible cause: The pen is not touching the tape. Adjust the height of the marker pen. Gently shake the box and pull the tape until a mark appears.

Problem: The adding machine tape does not move easily through the slots.

Possible cause: The slots are too small. Enlarge the slots to allow the tape to move freely.

Problem: The model works during the test, but when your friends run or jump, nothing happens.

Possible cause: The friends are not making strong enough vibrations. Have them jump up and down. If that doesn't work, have them move closer to the siesmograph.

Be sure to change only one variable at a time. Otherwise, you will not be able to determine which variable affected the results.

Experiment 2
Earthquake Simulation: Is the destruction greater at the epicenter?

Purpose/Hypothesis

In this experiment, you will create a simulated city and suburbs with buildings and houses. By locating different types of structures at various distances from the epicenter, you will determine the destructive power of an earthquake.

You probably have an educated guess about the outcome of this experiment based on what you already know about earthquakes. This

What Are the Variables?

Variables are anything that might affect the results of an experiment. Here are the main variables in this experiment:

- the size of the balloon, hence the amount of the simulated earthquake disturbance
- the positions of the buildings in the simulated city and suburb areas
- the height of the buildings
- the type of building construction
- the surface on which the buildings are constructed

In other words, the variables in this experiment are everything that might affect the amount of destruction. If you change more than one variable, you will not be able to tell which variable had the most effect on the seismograph recordings.

educated guess, or prediction, is your **hypothesis.** A hypothesis should explain these things:

- the topic of the experiment
- the **variable** you will change
- the variable you will measure
- what you expect to happen

A hypothesis should be brief, specific, and measurable. It must be something you can test through observation. Your experiment will prove or disprove whether your hypothesis is correct. Here is one possible hypothesis for this experiment: "Greater destruction occurs at the epicenter than at the outer limits of an earthquake."

In this case, the variable you will change is the distance from the simulated earthquake disturbance, and the variable you will measure is the amount of visible destruction of the structures in your simulated city and suburbs. If there is more destruction near the epicenter, you will know your hypothesis is correct.

experiment
CENTRAL

Level of Difficulty

Easy/moderate.

Materials Needed

- cardboard sheet, 24 inches x 24 inches (60 centimeters x 60 centimeters)
- 8 sheets of 8-1/2-inch x 11-inch (22 centimeter x 28 centimeter) paper
- marking pen
- 30 sugar cubes
- 8–10 spherical balloons
- adhesive tape
- 4 coffee cans
- ruler
- drawing compass
- safety pin

Approximate Budget

$3 for balloons and sugar cubes.

Timetable

1 hour or less.

Step-by-Step Instructions

1. Using tape, connect the edges of four sheets of paper to form a large rectangle—two sheets wide by two sheets long.

2. In the center of the rectangle, where the four corners join together, draw a small bullseye with the compass. Adjust the compass so the first circle has a 1-inch (2.5-centimeter) radius around the center of the bullseye. Continue to draw circles so that each is 1 inch

How to Experiment Safely

Use caution when blowing up and handling balloons. Ask an adult to help. Place the safety pin in fabric or cardboard when it is not being used. Discard the sugar cubes after you have used them.

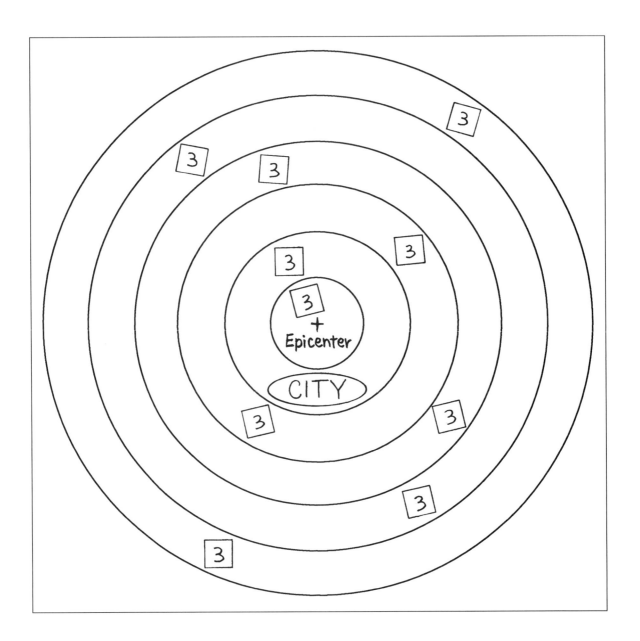

(2.5 centimeters) bigger in radius than the circle inside it. Mark the center of the bullseye X; this will be the epicenter. Label the paper "City."

3. Using the above illustration as a guide, randomly place ten sugar cubes on your City bullseye pattern. These represent city dwellings of three stories. Outline these cubes on the paper with your marking pen, and write 3, for three stories, in the center of the outlines.

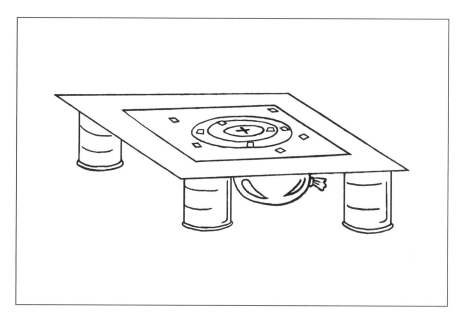

Steps 6 to 9: Set-up of simulated earthquake using City bullseye pattern.

4. Repeat steps 1 and 2 with the remaining pieces of paper, only this time label the paper "Suburb."

5. Randomly place ten sugar cubes on your Suburb bullseye pattern. Outline these cubes with your marking pen and mark 1 in the center of the outlines. These represent a rural or suburban area that has one-story homes.

6. Place the four coffee cans in a square pattern about 24 inches (61 centimeters) from each other.

7. Place the cardboard sheet on top of the coffee cans.

8. Blow up two balloons. Make them full, but small enough to fit under the cardboard sheet. Tape one to the center of the underside of the cardboard.

9. Place the City bullseye pattern on the cardboard. Try to position the epicenter mark directly over the spot where the balloon is taped.

10. Stack three sugar cubes on top of each other over each outline.

11. Using your safety pin, carefully pop the balloon.

12. Using a marking pen and ruler, mark and measure the new positions of the cubes with dotted lines.

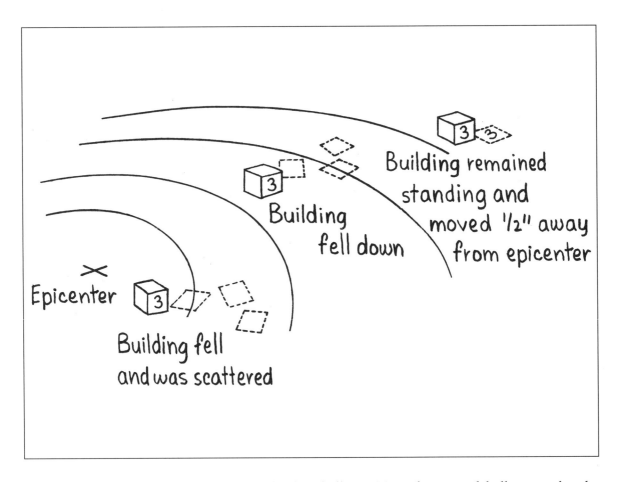

Epicenter

Building fell
and was scattered

Building
fell down

Building remained
standing and
moved '/2" away
from epicenter

Step 12: Record seismic movement from simulated earthquake.

13. Remove the broken balloon. Tape the second balloon under the center of the cardboard sheet and repeat steps 9 through 12 for the Suburb bullseye pattern. This time, place only one sugar cube over each outline.

Summary of Results

Compare the destruction on your two bullseye patterns. How did the simulated city compare to the suburb? Write up your results and describe the differences. Did your hypothesis hold true? Was the destruction near the epicenter greater in both cases?

Change the Variables

You can change the variables and repeat the experiment. For example, you can change the thickness of the cardboard to determine if the destruction increases or decreases. You can also change the height of the buildings. One interesting experiment might be to pick one of the

Troubleshooter's Guide

Your experiment may not have worked out as planned. Below is a problem that may arise during this experiment, a possible cause, and a way to solve the problem.

Problem: My balloon is not creating much damage.

Possible cause: The cardboard may be too thick and is absorbing the jolt. Try a thinner piece of cardboard. Also make sure the balloon is firmly attached to the cardboard.

three-story building outlines near the epicenter and place four stacks of three sugar cubes centered on the outline and arranged in a tight square so the stacks are touching. You can then compare the amount of damage of this type of building construction with a single three-story stack. Does a wider and broader base increase or decrease the amount of destruction?

Remember to change only one variable at a time or you will not be able to determine which variable affected the results.

Design Your Own Experiment

How to Select a Topic Relating to this Concept

Earth is dynamic and changing. Earthquakes, volcanoes, and tidal waves called **tsunamis** (pronounced SUE-nahm-ease; large waves of water caused by underwater earthquakes) are disastrous forces of nature that demonstrate Earth's motion. If you are fascinated with the power of these natural disasters, you can explore topics relating to earth science.

Major earthquakes are always reported in newspapers. You can look up major earthquakes in your local library. Newspaper accounts cover details such as seismic activity and the severity of the earthquake. One of the more recent ones in the United States took place in 1989 in San Francisco. Another took place in Turkey in 1999.

Steps in the Scientific Method

To do an original experiment, you need to plan carefully and think things through. Otherwise, you might not be sure what question you are answering, what you are or should be measuring, or what your findings prove or disprove.

Here are the steps in designing an experiment:

- State the purpose of—and the underlying question behind—the experiment you propose to do.
- Recognize the variables involved, and select one that will help you answer the question at hand.
- State a testable hypothesis, an educated guess about the answer to your question.
- Decide how to change the variable you selected.
- Decide how to measure the results.

Recording Data and Summarizing Results

Your experiment can be useful to others studying the same topic. When designing your experiment, develop a simple method to record your data. This method should be simple and clear enough so that others who want to do the experiment can follow it.

Your final results should be summarized and put into simple graphs, tables, and charts to display the outcome of your experiment.

Related Projects

Building an actual model of a city, town, or region that can be affected by a simulated earthquake is another way to understand the dynamics of a real earthquake.

For More Information

Bolt, Bruce A. *Earthquakes and Geological Discovery.* New York: Scientific American Library, 1997. ❖ Offers geological facts and photos about earthquakes.

Smith, Bruce, and David McKay. *Geology Projects For Young Scientists.* New York: Franklin Watts, 1992. ❖ Describes earthquake experiments and the geological background of why earthquakes occur.

Van Rose, Susanna. *Volcano and Earthquake.* New York: Alfred A. Knopf, 1992. ❖ Explains the cause and effect of volcanoes and earthquakes. Includes chapters on seismic waves and early measuring devices, including Chang Heng's seismoscope.

chapter 12

Eclipses

Imagine living in ancient times. You stroll down a dirt road leading to a favorite temple. It is a nice day, but without warning, the sky starts to get dark. The Sun looks strange and, gradually, something huge blocks it out, although a bright ring can be seen around its edge.

We now know that this phenomenon is a **solar eclipse**. An **eclipse** occurs when one **celestial body** passes in front of another, partly or completely cutting off our view of it. Today, we would get advance information through newspapers and magazines or by news reports on television or radio if a major eclipse was expected. To most ancient people, who had no explanations for the darkness, an eclipse was terrifying.

Close encounters in the sky

In the eighth century B.C., Babylonian scholars began systematically observing and writing down celestial phenonema, as they studied **astronomy**. These scholars watched the motion of the planets and noticed that sometimes two planets came close together. Sometimes the Moon passed in front of the Sun. Sometimes Earth's shadow fell on the Moon. After studying these phenomena for many years, they identified certain experiences as occurring in **cycles.** They also developed mathematical formulas involving time and distances that helped them to predict eclipses.

Thales of Miletus (624–546 B.C.) was a Greek philosopher who may have learned astronomical methods from the Babylonian scholars. Thales accurately predicted a solar eclipse on May 28, 585 B.C.—probably the earliest, most public eclipse prediction. The term *eclipse* comes

The Moon completely blocks out our view of the Sun during a solar eclipse. (Photo Researchers Inc. Reproduced by permission.)

from the Greek words meaning "to leave out," because when one occurred, either the Sun or the Moon was "left out." In fact, the theory that Earth was a sphere began getting attention around this time because observers noticed that Earth's shadow on the Moon during eclipses was always circular.

The first eclipse to interest a significant number of astronomers took place on April 22, 1715. The shadow of the eclipse fell across Great Britain and parts of Europe. English astronomer Edmond Halley (1656–1742) plotted its path and prepared maps enabling many to watch its course.

Celestial line-up

The two most commonly known eclipses are solar and lunar. Earth revolves around the Sun. The Moon revolves around Earth. The Moon takes a month to complete a revolution; Earth takes a year. Sometimes these three bodies end up in a straight line and cause an eclipse.

Two conditions have to be met for a **total solar eclipse**—one in which our view of the Sun is completely blocked. The Sun, Moon, and

Earth must lie in a perfectly straight line, and the Moon must be a certain distance from Earth to cover the Sun. When these conditions are met, the Moon totally blocks our view of the Sun for a period of about seven minutes. If the Moon is too far away from Earth, or if it is not exactly aligned between Earth and the Sun, it will only partially block the Sun, causing a **partial solar eclipse.**

For a **total lunar eclipse,** the Sun, Earth, and Moon must lie in a perfectly straight line. Did you catch the difference? In this case, Earth is in the middle, not the Moon. Earth's shadow across the Moon is what causes the darkness. Lunar eclipses can happen only during a full Moon, when Earth's dark side faces the Moon's bright side. In this position, Earth casts a shadow, causing the Moon to darken.

Celestial fireworks

The bright ring you might see around the Sun during a solar eclipse is the **corona,** the Sun's outermost layer, which appears to be a pearly color. The red plumes that shoot out around this ring are called **prominences.**

Red light waves from the Sun cause the Moon to turn a reddish color during a lunar eclipse. (Photo Researchers Inc. Reproduced by permission.)

(W)**ords to Know**

Hypothesis:
An idea in the form of a statement that can be tested by observation and/or experiment.

Lunar eclipse:
An eclipse that occurs when Earth passes between the Sun and the Moon, casting a shadow on the Moon.

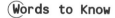

In 1869, British astronomer Joseph Norman Lockyer became the first person to observe solar prominences in the daytime. (Photo Researchers Inc. Reproduced by permission.)

Like fireworks, these streams of glowing gas shoot out from the Sun and extend many miles into space. No wonder ancient people were terrified. Lunar eclipses have a colorful side also. They can make the Moon turn red. This reddish color is actually an accumulation of light waves from the Sun.

By constructing models that simulate eclipses, we can better understand the extraordinary processes that cause them.

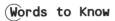

Project 1
Simulating Solar and Lunar Eclipses

Purpose
This project will create a model that demonstrates a solar and lunar eclipse. By adjusting the alignment and distances of the model Sun, Moon, and Earth, you should be able to demonstrate both partial and total eclipses.

Level of Difficulty
Easy/moderate. (The assembly and principles are not difficult, but it takes patience to adjust the objects to get the desired effect.)

Materials Needed
*   2 Styrofoam balls, one ball 2 inches (5 centimeters) and one 0.5 inch (1.25 centimeters) in diameter
*   two 4-inch (10-centimeter) Styrofoam squares
*   small table lamp (measuring 12 inches in height) with no lamp shade and a 40-watt bulb

How to Work Safely

Use caution when handling the lamp. Do not touch or move it until it has cooled for at least 5 minutes.

- 2 wooden dowels (as long as the height of the lamp from its base to the middle of the bulb)
- ruler

Approximate Budget

$3 for the Styrofoam pieces and the dowels.

Timetable

Less than 1 hour.

Step-by-Step Instructions

1. Poke each dowel into the center of a Styrofoam square.

2. Place the small Styrofoam ball, representing the Moon, onto one dowel.

3. Place the large Styrofoam ball, representing Earth, onto the other dowel.

4. Place the lamp on a sturdy table and plug it in. Turn it on.

5. Here is the challenge! Place the Sun (lamp), Earth (large ball), and Moon (small ball) on a flat surface in perfect alignment to create a solar and lunar eclipse.

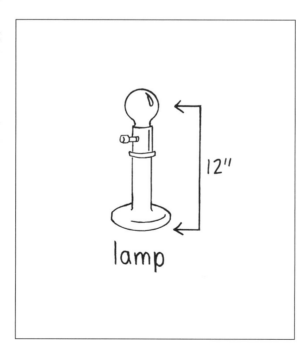

Lamp without shade and measured distance of 12 inches.

LEFT: Step 5:
Solar eclipse set-up.

RIGHT: Step 5:
Lunar eclipse set-up.

Follow the diagrams illustrated.

Summary of Results

Make a diagram of your experiment, measuring and marking the distances and height of the experiment parts for others to see and try. Through the shadows you created with the lamp, were you able to create full eclipses or only partial eclipses?

Troubleshooter's Guide

Here is a problem that may arise during this project, a possible cause, and a way to remedy the problem.

Problem: You cannot get the shadow to cover the entire object to create the "eclipse."

Possible cause: Your alignment may be off. Make sure you line up the objects on the same level.

experiment
CENTRAL

Project 2
Phases of the Moon: What does each phase look like?

Purpose
In this project, you will create models of the changes in the illuminated Moon surface as the Moon revolves around Earth. These changes are called **phases.** You will create diagrams called sun prints representing these Moon phases.

Level of Difficulty
Easy/moderate.

Materials Needed
- 8 sheets of dark blue construction paper, 8 1/2 x 11 inches (21.5 x 28 centimeters)
- 8 sheets of black construction paper, 8 1/2 x 11 inches (21.5 x 28 centimeters)
- adhesive tape
- marker
- 30 x 30-inch (75 x 75-centimeter) board
- sunny day
- scissors
- drawing compass

Approximate Budget
$5 for paper supplies.

Timetable
Approximately 1 hour to set up the model and a whole day for the sun prints to mature.

How to Work Safely
Use caution with the compass and scissors.

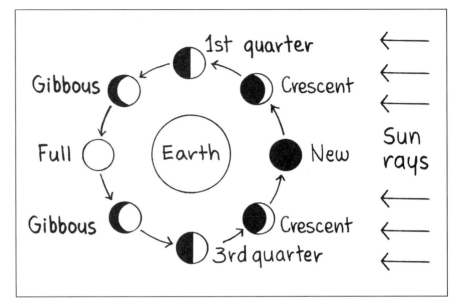

Step 5: Set-up for record-ing phases of the moon.

Step-by-Step Instructions

1. Use the compass to draw a 7-inch (18-centimeter) diameter circle on eight sheets of blue construction paper.

2. Draw an 8-inch (20-centimeter) diameter circle on eight sheets of black construction paper.

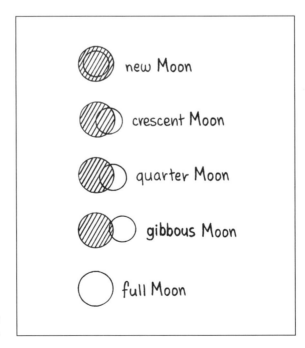

Steps 6 to 8: Completed sun prints.

3. Cut out the circles.

4. Tape eight blue circles onto the board in a circle.

5. Mark the board as shown in the diagram illustrated above.

6. Place the black circles over the blue circles to show:

    new Moon
    crescent Moon
    first-quarter Moon
    gibbous Moon
    full Moon
    gibbous Moon

Troubleshooter's Guide

Here is a problem that may arise during this project, a possible cause, and a way to remedy the problem.

Problem: The sun prints are not forming.

Possible cause: They have not had enough time. Give the sun prints 2 days, for 8 hours each day, in full sunlight.

third-quarter Moon
crescent Moon

7. Leave the board in a sunny location for at least 8 hours.
8. Take the black paper off after 8 hours and examine the results.
9. Highlight the lightened areas or boundaries with the marker.

Note: the darker blue areas that were covered are the shaded part of the Moon we cannot see.

Summary of Results

Label the board and write a brief description for each Moon phase, that is, how it was caused and what it looks like.

 # Design Your Own Experiment

How to Select a Topic Relating to This Concept

Astronomy is a fascinating field of study, with topics such as meteor/meteorites, telescopes, space travel, and stars. Read your local paper to find out about upcoming events in the sky. Then research who saw the phenomena first and when and how theories developed.

Check the For More Information section and talk with your science teacher or school or community media specialist to start gathering information on eclipse questions that interest you.

Steps in the Scientific Method

To do an original experiment, you need to plan carefully and think things through. Otherwise, you might not be sure what question you

experiment
CENTRAL

are answering, what you are or should be measuring, or what your findings prove or disprove.

Here are the steps in designing an experiment:

- State the purpose of—and the underlying question behind—the experiment you propose to do.
- Recognize the **variables** involved, and select one that will help you answer the question at hand.
- State a testable **hypothesis,** an educated guess about the answer to your question.
- Decide how to change the variable you selected.
- Decide how to measure your results.

Recording Data and Summarizing the Results

When performing an experiment, it is important to keep your data organized in tables. Your information needs to be analyzed and presented in a visual manner. Graphs, drawings, or pictures of events are great tools for displaying your data.

Related Projects

Creating models like these are always fun and interesting. However, creating a mini-instrument, such as a telescope with lenses and cardboard, might be useful. Ask a teacher or your parents for help.

For More Information

Aronson, Billy. *Eclipses: Nature's Blackouts.* New York: Franklin Watts, 1996. ❖ Explains what causes eclipses of the Sun and Moon and describes how they have been viewed and studied at different times in history.

Branley, Franklyn M. *Eclipse: Darkness in Daytime.* New York: HarperCollins Publishers, 1988. ❖ Provides an overview of this scientific phenonemon and famous historical eclipses.

budget index

Under $5

[Annual Growth] What can be learned from the
growth patterns of trees? **1:**24

[Chemical Properties] What happens when white glue
and borax mix? **1:**79

[Density and Buoyancy] Does water pressure
affect buoyancy? **1:**132

[Earthquakes] How can movement of Earth's crust
be measured? **1:**162

[Earthquakes] Is the destruction greater at
the epicenter? **1:**167

[Eclipses] Simulating solar and lunar eclipses. **1:**178

[Flight] Will it fly high? **2:**256

[Germination] How fast can seeds grow? **2:**270

[Gravity] How fast can a balance be made? **2:**285

[Greenhouse Effect] What happens when fossil
fuels burn? **2:**300

[Heat] How does heat move through liquids? **2:**334

[Heat] Which solid materials are the best
conductors of heat? **2:**327

[Microorganisms] What is the best way to
grow penicillin? **3:**389

[Mixtures and Solutions] Can filtration and
evaporation determine whether mixtures are
suspensions or solutions? **3:**406

[Nutrition] Which foods contain carbohydrates and fats? **3:**423

Bold type indicates volume number.

budget index

[Optics and Optical Illusions] Can the eye be fooled? 3:437

[Osmosis and Diffusion] Will a bag of salt water
draw in fresh water? 3:453

[Oxidation-Reduction] How will acid affect
dirty pennies? 3:463

[pH] What is the pH of household chemicals? 3:481

[Potential and Kinetic Energy] How does the height
of an object affect its potential energy? 3:512

[Rocks and Minerals] Is it igneous, sedimentary,
or metamorphic? 3:534

[Scientific Method] Do fruit flies appear out of thin air? 3:568

[Stars] Can a planet be followed? 4:609

[Static Electricity] Does nylon or wool create a stronger
static electric charge? 4:625

[Tropisms] Will plants follow a maze to reach light? 4:650

[Weather Forecasting] How can air pressure
be measured? 4:769

$5—$10

[Acid Rain] How does acid rain affect brine shrimp? 1:4

[Acid Rain] How does acid rain affect plant growth? 1:10

[Annual Growth] What can be learned from the
environment by observing lichens? 1:28

[Cells] What are the cell differences between monocot
and dicot plants? 1:55

[Cells] What are the differences between a multicellular
organism and a unicellular organism? 1:52

[Chemical Energy] Is the chemical reaction exothermic,
endothermic, or neither? 1:64

[Chemical Properties] What happens when mineral oil,
water, and iodine mix? 1:84

[Chlorophyll] Can pigments be separated? 1:93

[Composting/Landfills] Using organic material to
grow plants 1:115

[Composting/Landfills] What effect do the
microorganisms in soil have on the
decomposition process? 1:109

[Density and Buoyancy] Can a scale of relative density
predict whether one material floats on another? 1:126

[Eclipses] What does each phase look like? 1:181

[Enzymes] Which enzyme breaks down
hydrogen peroxide? 2:220
[Erosion] Does soil type affect the amount of water
that runs off a hillside? 2:233
[Erosion] How do plants affect the rate of soil erosion? 2:239
[Flight] How can a glider be made to fly higher? 2:252
[Germination] What temperatures encourage and
discourage germination? 2:266
[Gravity] How fast do different objects fall? 2:280
[Greenhouse Effect] How much will the temperature
rise inside a greenhouse? 2:294
[Groundwater Aquifers] How can it be cleaned? 2:316
[Magnetism] Does the strength of an electromagnet
increase with greater current? 3:379
[Magnetism] How do heat, cold, jarring, and rubbing
affect the magnetism of a nail? 3:373
[Microorganisms] Growing microorganisms in a
petri dish 3:394
[Osmosis and Diffusion] Is a plastic bag a
semipermeable membrane? 3:448
[Oxidation-Reduction] How is rust produced? 3:469
[pH] What is required to change a substance from an
acid or a base into a neutral solution? 3:486
[Photosynthesis] How does light affect plant growth? 3:496
[Properties of Water] Can the cohesive force of surface
tension in water support an object denser than water? 4:701
[Rivers of Water] Does the stream meander? 4:721
[Rocks and Minerals] What kind of mineral is it? 3:531
[Salinity] How can salinity be measured? 3:544
[Salinity] How to make a standard for measuring density 3:550
[Scientific Method] What are the mystery powders? 3:562
[Sound] How does the length of a vibrating string
affect the sound it produces? 4:592
[Sound] How does the thickness of a vibrating string
affect sound? 4:596
[Static Electricity] Which objects are electrically charged? 4:619
[Vegetative Propagation] How do potatoes reproduce
vegetatively? 4:676
[Volcanoes] Can a volcanic eruption be detected? 4:691

budget index

Bold type indicates volume number.

[Water Cycle] How does surface area affect the rate
    of evaporation? 4:736

[Weather] Measuring wind speed with a homemade
    anemometer 4:748

[Weather] Will a drop in air temperature cause a cloud
    to form? 4:752

[Weather Forecasting] When will dew form? 4:764

$11—$15

[Enzymes] Does papain speed up the aging process? 2:224

[Mixtures and Solutions] Can colloids be distinguished
    from suspension using the Tyndall effect? 3:411

[Properties of Water] How much weight is required to
    break the adhesive force between an object and water? 4:705

[Rivers of Water] How does a river make a trench? 4:717

[Solar Energy] Will seedlings grow bigger in a greenhouse? ... 4:579

[Structures and Shapes] How does the vertical height
    of a beam affect its rigidity? 4:640

[Structures and Shapes] Which is strongest? 4:637

[Volcanoes] Will it blow its top? 4:687

[Water Cycle] How does temperature affect the rate
    of evaporation? 4:731

$16—$20

[Dissolved Oxygen] How does a decrease in the dissolved
    oxygen level affect the breathing rate of goldfish? 1:149

[Electromagnetism] How can a magnetic field be created
    and detected? 2:206

[Electromagnetism] How can an electromagnet
    be created? 2:210

[Groundwater Aquifers] How do they become polluted? 2:311

[Nutrition] Which foods contain proteins and salts? 3:426

[Optics and Optical Illusions] What is the focal length
    of a lens? 3:432

[Potential and Kinetic Energy] Build a roller coaster 3:517

[Stars] Where is Polaris? 4:606

[Tropisms] Will plant roots turn toward the pull
    of gravity? 4:655

[Vegetative Propagation] How do auxins affect
    plant growth? 4:668

$21—$25

[Biomes] Building a desert biome **1**:42

[Biomes] Building a temperate forest biome **1**:39

[Chemical Energy] Determining whether various
chemical reactions are exothermic or endothermic **1**:68

[Dissolved Oxygen] How does the amount of decaying
matter affect the level of dissolved oxygen in water? **1**:144

[Photosynthesis] How does the intensity of light affect
plant growth? **3**:501

[Properties of Light] Which objects glow under
black light? **2**:360

[Solar energy] Will sunlight make a motor run? **4**:584

budget index

$26—$30

[Chlorophyll] Do plants grow differently in different
colors of light? **1**:98

[Electricity] Can a series of homemade electric
cells form a "pile" strong enough to match the
voltage of a D-cell battery? **2**:193

[Electricity] Do some solutions conduct electricity
better than others? **2**:188

$31—$35

[Life Cycles] Does temperature affect the rate at
which tadpoles change into frogs? **2**:343

[Life Cycles] How does food supply affect the growth
rate of grasshoppers or crickets? **2**:349

[Properties of Light] Making a rainbow **2**:363

Bold type indicates volume number.

level of difficulty index

Easy

Easy means that the average student should easily be able to complete the tasks outlined in the project/experiment, and that the time spent on the project is not overly restrictive.

[Electromagnetism] How can an electromagnet
be created? 2:210

[Flight] How can a glider be made to fly higher? 2:252

[Flight] Will it fly high? 2:256

[Nutrition] Which foods contain carbohydrates and fats? 3:423

[Osmosis and Diffusion] Will a bag of salt water draw
in fresh water? 3:453

[Potential and Kinetic Energy] How does the height of
an object affect its potential energy? 3:512

[Rivers of Water] Does the stream meander? 4:721

[Sound] How does the length of a vibrating string affect
the sound it produces? 4:592

[Sound] How does the thickness of a vibrating string
affect sound? 4:596

[Volcanoes] Can a volcanic eruption be detected? 4:691

[Water Cycle] How does surface area affect the rate
of evaporation? 4:736

[Water Cycle] How does temperature affect the rate
of evaporation? 4:731

[Weather Forecasting] How can air pressure be measured? 4:769

[Weather Forecasting] When will dew form? 4:764

Bold type indicates volume number.

Easy/Moderate

Easy/Moderate means that the average student should have little trouble completing the tasks outlined in the project/experiment, and that the time spent on the project is not overly restrictive.

[Chemical Properties] What happens when mineral oil, water, and iodine mix? 1:84

[Chemical Properties] What happens when white glue and borax mix? 1:79

[Composting/Landfills] What effect do the microorganisms in soil have on the decomposition process? 1:109

[Dissolved Oxygen] How does a decrease in the dissolved oxygen level affect the breathing rate of goldfish? 1:149

[Dissolved Oxygen] How does the amount of decaying matter affect the level of dissolved oxygen in water? 1:144

[Earthquakes] Is the destruction greater at the epicenter? 1:167

[Eclipses] Simulating solar and lunar eclipses 1:178

[Eclipses] What does each phase look like? 1:181

[Electricity] Can a series of homemade electric cells form a "pile" strong enough to match the voltage of a D-cell battery? 2:193

[Enzymes] Does papain speed up the aging process? 2:224

[Enzymes] Which enzyme breaks down hydrogen peroxide? 2:220

[Germination] How fast can seeds grow? 2:270

[Germination] What temperatures encourage and discourage germination? 2:266

[Gravity] How fast do different objects fall? 2:280

[Greenhouse Effect] How much will the temperature rise inside a greenhouse? 2:294

[Heat] How does heat move through liquids? 2:334

[Magnetism] Does the strength of an electromagnet increase with greater current? 3:379

[Magnetism] How do heat, cold, jarring, and rubbing affect the magnetism of a nail? 3:373

[Microorganisms] What is the best way to grow penicillin? ... 3:389

[Oxidation-Reduction] How is rust produced? 3:469

[Properties of Light] Making a rainbow 2:363

[Properties of Light] Which objects glow under black light? 2:360

[Properties of Water] Can the cohesive force of surface
tension in water support an object denser than water? 4:701
[Properties of Water] How much weight is required to
break the adhesive force between an object and water? 4:705
[Scientific Method] Do fruit flies appear out of thin air? 3:568
[Scientific Method] What are the mystery powders? 3:562
[Solar Energy] Will seedlings grow bigger in
a greenhouse? 4:579
[Solar energy] Will sunlight make a motor run? 4:584
[Static Electricity] Does nylon or wool create a stronger
static electric charge? 4:625
[Static Electricity] Which objects are electrically charged? 4:619
[Structures and Shapes] Which is strongest? 4:637
[Weather] Measuring wind speed with a
homemade anemometer 4:748
[Weather] Will a drop in air temperature cause a
cloud to form? 4:752

Moderate

Moderate means that the average student should find tasks outlined in the project/experiment challenging but not difficult, and that the time spent project/experiment may be more extensive.

[Acid Rain] How does acid rain affect brine shrimp? 1:4
[Acid Rain] How does acid rain affect plant growth? 1:10
[Annual Growth] What can be learned from the
environment by observing lichens? 1:28
[Annual Growth] What can be learned from the growth
patterns of trees? 1:24
[Biomes] Building a temperate forest biome 1:39
[Chemical Energy] Determining whether various
chemical reactions are exothermic or endothermic 1:68
[Chemical Energy] Is the chemical reaction
exothermic, endothermic, or neither? 1:64
[Chlorophyll] Can pigments be separated? 1:93
[Chlorophyll] Do plants grow differently in different
colors of light? 1:98
[Composting/Landfills] Using organic material to
grow plants 1:115

Bold type indicates volume number.

level of difficulty index

[Density and Buoyancy] Can a scale of relative density predict whether one material floats on another? 1:126

[Density and Buoyancy] Does water pressure affect buoyancy? 1:132

[Earthquakes] How can movement of Earth's crust be measured? 1:162

[Electricity] Do some solutions conduct electricity better than others? 2:188

[Electromagnetism] How can a magnetic field be created and detected? 2:206

[Erosion] Does soil type affect the amount of water that runs off a hillside? 2:233

[Erosion] How do plants affect the rate of soil erosion? 2:239

[Gravity] How fast can a balance be made? 2:285

[Greenhouse Effect] What happens when fossil fuels burn? 2:300

[Groundwater Aquifers] How can it be cleaned? 2:316

[Groundwater Aquifers] How do they become polluted? 2:311

[Microorganisms] Growing microorganisms in a petri dish 3:394

[Mixtures and Solutions] Can colloids be distinguished from suspension using the Tyndall effect? 3:411

[Mixtures and Solutions] Can filtration and evaporation determine whether mixtures are suspensions or solutions? 3:406

[Nutrition] Which foods contain proteins and salts? 3:426

[Optics and Optical Illusions] What is the focal length of a lens? 3:432

[Osmosis and Diffusion] Is a plastic bag a semipermeable membrane? 3:448

[Oxidation-Reduction] How will acid affect dirty pennies? 3:463

[Photosynthesis] How does light affect plant growth? 3:496

[Photosynthesis] How does the intensity of light affect plant growth? 3:501

[Potential and Kinetic Energy] Build a roller coaster 3:517

[Rivers of Water] How does a river make a trench? 4:717

[Rocks and Minerals] Is it igneous, sedimentary, or metamorphic? 3:534

[Salinity] How to make a standard for measuring density 3:550

[Stars] Can a planet be followed? 4:609

experiment
CENTRAL

[Stars] Where is Polaris? 4:606

[Structures and Shapes] How does the vertical height
of a beam affect its rigidity? 4:640

[Tropisms] Will plant roots turn toward the
pull of gravity? 4:655

[Tropisms] Will plants follow a maze to reach light? 4:650

[Vegetative Propagation] How do auxins affect
plant growth? 4:668

[Vegetative Propagation] How do potatoes
reproduce vegetatively? 4:676

[Volcanoes] Will it blow its top? 4:687

Moderate/Difficult

Moderate/Difficult means that the average student should find tasks out-lined in the project/experiment challenging, and that the time spent on the project/experiment may be more extensive.

[Biomes] Building a desert biome 1:42

[Cells] What are the cell differences between monocot
and dicot plants? 1:55

[Cells] What are the differences between a multicellular
organism and a unicellular organism? 1:52

[Heat] Which solid materials are the best conductors
of heat? 2:327

[Rocks and Minerals] What kind of mineral is it? 3:531

[Salinity] How can salinity be measured? 3:544

Difficult

Difficult means that the average student will probably find the tasks out-lined in the project/experiment mentally and physically challenging, and that the time spent on the project/experiment will be more extensive.

[Life Cycles] Does temperature affect the rate at which
tadpoles change into frogs? 2:343

[Life Cycles] How does food supply affect the growth
rate of grasshoppers or crickets? 2:349

[Optics and Optical Illusions] Can the eye be fooled? 3:437

[pH] What is required to change a substance from an
acid or a base into a neutral solution? 3:486

[pH] What is the pH of household chemicals? 3:481

Bold type indicates volume number.

timetable index

Less than 15 minutes

[Greenhouse Effect] What happens when fossil
fuels burn? 2:300
[Properties of Light] Which objects glow under
black light? 2:360

20 minutes

[Density and Buoyancy] Does water pressure
affect buoyancy? 1:132
[Electricity] Can a series of homemade electric
cells form a "pile" strong enough to match the
voltage of a D-cell battery? 2:193
[Enzymes] Which enzyme breaks down
hydrogen peroxide? 2:220
[Flight] Will it fly high? 2:256
[Gravity] How fast do different objects fall? 2:280
[Heat] How does heat move through liquids? 2:334
[Magnetism] Does the strength of an electromagnet
increase with greater current? 3:379
[Properties of Water] Can the cohesive force of surface
tension in water support an object denser than water? 4:701
[Rocks and Minerals] What kind of mineral is it? 3:531
[Static Electricity] Does nylon or wool create a
stronger static electric charge? 4:625
[Weather] Measuring wind speed with a
homemade anemometer 4:748

Bold type indicates
volume number.

30-45 minutes

[Annual Growth] What can be learned from the growth
   patterns of trees? 1:24

[Chemical Energy] Is the chemical reaction exothermic,
   endothermic, or neither? 1:64

[Chemical Properties] What happens when mineral oil,
   water, and iodine mix? 1:84

[Flight] How can a glider be made to fly higher? 2:252

[Gravity] How fast can a balance be made? 2:285

[Heat] Which solid materials are the best conductors
   of heat? 2:327

[Magnetism] How do heat, cold, jarring, and rubbing
   affect the magnetism of a nail? 3:373

[Properties of Light] Making a rainbow 2:363

[Rivers of Water] Does the stream meander? 4:721

[Salinity] How to make a standard for measuring density. 3:550

[Scientific Method] What are the mystery powders? 3:562

[Solar energy] Will sunlight make a motor run? 4:584

[Static Electricity] Which objects are electrically charged? 4:619

[Structures and Shapes] How does the vertical height
   of a beam affect its rigidity? 4:640

[Structures and Shapes] Which is strongest? 4:637

[Volcanoes] Can a volcanic eruption be detected? 4:691

[Weather Forecasting] When will dew form? 4:764

1 hour

[Cells] What are the cell differences between monocot
   and dicot plants? 1:55

[Cells] What are the differences between a multicellular
   organism and a unicellular organism? 1:52

[Chemical Energy] Determining whether various
   chemical reactions are exothermic or endothermic. 1:68

[Chemical Properties] What happens when white glue
   and borax mix? 1:79

[Density and Buoyancy] Can a scale of relative density
   predict whether one material floats on another? 1:126

[Earthquakes] How can movement of Earth's crust
   be measured? 1:162

[Earthquakes] Is the destruction greater at the epicenter? 1:167

[Eclipses] Simulating solar and lunar eclipses **1**:178

[Electricity] Do some solutions conduct electricity
better than others? **2**:188

[Mixtures and Solutions] Can colloids be distinguished
from suspension using the Tyndall effect? **3**:411

[Nutrition] Which foods contain carbohydrates and fats? **3**:423

[Nutrition] Which foods contain proteins and salts? **3**:426

[pH] What is required to change a substance from an
acid or a base into a neutral solution? **3**:486

[pH] What is the pH of household chemicals? **3**:481

[Potential and Kinetic Energy] How does the height
of an object affect its potential energy? **3**:512

[Rocks and Minerals] Is it igneous, sedimentary,
or metamorphic? **3**:534

[Salinity] How can salinity be measured? **3**:544

[Sound] How does the length of a vibrating string
affect the sound it produces? **4**:592

[Sound] How does the thickness of a vibrating string
affect sound? **4**:596

[Weather] Will a drop in air temperature cause a
cloud to form? **4**:752

2 hours

[Chlorophyll] Can pigments be separated? **1**:93

[Electromagnetism] How can a magnetic field be created
and detected? **2**:206

[Electromagnetism] How can an electromagnet
be created? **2**:210

[Groundwater Aquifers] How can it be cleaned? **2**:316

[Groundwater Aquifers] How do they become polluted? **2**:311

[Optics and Optical Illusions] What is the focal length
of a lens? **3**:432

[Oxidation-Reduction] How will acid affect
dirty pennies? **3**:463

[Potential and Kinetic Energy] Build a roller coaster **3**:517

[Properties of Water] How much weight is required to
break the adhesive force between an object and water? **4**:705

[Stars] Where is Polaris? **4**:606

Bold type indicates volume number.

timetable index

3 hours

[Annual Growth] What can be learned from the environment by observing lichens? 1:28

[Erosion] Does soil type affect the amount of water that runs off a hillside? 2:233

[Mixtures and Solutions] Can filtration and evaporation determine whether mixtures are suspensions or solutions? 3:406

[Volcanoes] Will it blow its top? 4:687

6 hours

[Dissolved Oxygen] How does a decrease in the dissolved oxygen level affect the breathing rate of goldfish? 1:149

1 day

[Eclipses] What does each phase look like? 1:181

[Enzymes] Does papain speed up the aging process? 2:224

[Osmosis and Diffusion] Will a bag of salt water draw in fresh water? 3:453

[Water Cycle] How does temperature affect the rate of evaporation? 4:731

2 days

[Osmosis and Diffusion] Is a plastic bag a semipermeable membrane? 3:448

3 days

[Oxidation-Reduction] How is rust produced? 3:469

1 week

[Acid Rain] How does acid rain affect brine shrimp? 1:4

[Dissolved Oxygen] How does the amount of decaying matter affect the level of dissolved oxygen in water? 1:144

[Greenhouse Effect] How much will the temperature rise inside a greenhouse? 2:294

[Water Cycle] How does surface area affect the rate of evaporation? 4:736

2 weeks

[Acid Rain] How does acid rain affect plant growth? 1:10

[Erosion] How do plants affect the rate of soil erosion? 2:239

[Germination] How fast can seeds grow? 2:270

[Germination] What temperatures encourage and
    discourage germination? 2:266

[Microorganisms] Growing microorganisms in a
    petri dish 3:394

[Microorganisms] What is the best way to
    grow penicillin? 3:389

[Scientific Method] Do fruit flies appear out of thin air? 3:568

[Weather Forecasting] How can air pressure be measured? 4:769

3-4 weeks

[Life Cycles] Does temperature affect the rate at
    which tadpoles change into frogs? 2:343

[Life Cycles] How does food supply affect the growth
    rate of grasshoppers or crickets? 2:349

[Photosynthesis] How does light affect plant growth? 3:496

[Photosynthesis] How does the intensity of light affect
    plant growth? 3:501

[Rivers of Water] How does a river make a trench? 4:717

[Solar Energy] Will seedlings grow bigger in
    a greenhouse? 4:579

[Stars] Can a planet be followed? 4:609

[Tropisms] Will plant roots turn toward the pull
    of gravity? 4:655

[Tropisms] Will plants follow a maze to reach light? 4:650

[Vegetative Propagation] How do auxins affect
    plant growth? 4:668

[Vegetative Propagation] How do potatoes
    reproduce vegetatively? 4:676

2 months

[Chlorophyll] Do plants grow differently in different
    colors of light? 1:98

4 months

[Composting/Landfills] Using organic material to
    grow plants 1:115

[Composting/Landfills] What effect do the microorganisms
    in soil have on the decomposition process? 1:109

Bold type indicates volume number.

6 months

[Biomes] Building a desert biome 1:42

[Biomes] Building a temperate forest biome 1:39

general index

A

Abscission **1**: 92
Acceleration **2**: 278
Acid **1**: 1, 76, **3**: 477
Acid rain **1**: 1-18, **3**: 479, 480 [ill.]
Acoustics **4**: 591
Active solar energy system **4**: 576
Adhesion **4**: 697, 698 [ill.]
Aeration **2**: 316
Aerobic **1**: 108
Air pollution **1**: 2,
Algae **1**: 23, 143 [ill.]
Alignment **3**: 370
Alkaline **1**: 1
Amine **3**: 420
Ampere, Andre-Marie **2**: 185, 186 [ill.]
Amphibian **1**: 2, **2**: 342
Amplitude **4**: 589
Anaerobic **1**: 108
Andromeda Galaxy **2**: 278 [ill.]
Anemometer **4**: 746, 749 [ill.], 760 [ill.]
Animalcules **3**: 387
Annual growth **1**: 19-34
Anthocyanin **1**: 92
Antibodies **3**: 422
Aquifer **2**: 307, **4**: 729
Arch **4**: 634
Archimedes **1**: 125 [ill.]
Arrhenius, Svante **2**: 291
Artesian well **2**: 308

Artwork from Pompeii **4**: 686 [ill.]
Asexual reproduction **4**: 665
Astronomers **4**: 603
Astronomy **4**: 603
Atmosphere **2**: 291
Atmospheric pressure **4**: 745
Atoms **1**: 61, **2**: 203, **3**: 461, **4**: 615
Autotrophs **1**: 23
Auxins **4**: 648 [ill.], 666, 668 [ill.]

B

Bacteria **3**: 387
Barometer **4**: 762
Base **1**: 1, 76, **3**: 477
Batteries **3**: 462 [ill.]
Beam **4**: 635
Bean vine **4**: 649 [ill.]
Beriberi **3**: 420
Bernoulli, Daniel **2**: 250
Biochemical oxygen demand
    (BOD_5) **1**: 141
Biodegradable **1**: 108
Biomes **1**: 35-47
Bloodstream **3**: 446 [ill.]
Bond **1**: 61
Braided rivers **4**: 715
Bread dough **2**: 219 [ill.]
Bridge **4**: 636 [ill.]
Brine shrimp **1**: 4,

Bold type indicates volume number; [ill.] indicates illustration or photograph.

general index

Buoyancy **1:** 123-138, 124 [ill.], **3:** 545
Butterfly **2:** 342 [ill.]
By-products **2:** 300

C

Camera **4:** 607
Capillary action **4:** 699
Carbohydrates **3:** 421
Carbon dioxide **1:** 3
Carbon monoxide **1:** 3,
Carnivore **4:** 668
Carotene **1:** 92, **3:** 494
Catalysts **2:** 217
Caterpillar **2:** 341 [ill.]
Celestial bodies **1:** 175
Cells **1:** 49-59
Cell membrane **1:** 51
Centrifuge **3:** 404
Channel **4:** 715
Chanute, Octave **2:** 251
Cheese curd **3:** 390 [ill.]
Chemicals **1:** 1,
Chemical energy **1:** 61-74, **3:** 509
Chemical properties **1:** 75
Chemical reaction **1:** 61, 75, 77 [ill.]
Chlorophyll **1:** 91-104, 265, **3:** 493
Chloroplasts **1:** 52, 91, **3:** 493
Chromatography **1:** 93
Cleavage **3:** 533
Climate **4:** 745
Clouds **4:** 748 [ill.]
Coagulation **2:** 316, **3:** 405
Cohesion **4:** 697, 698 [ill.]
Colloid **3:** 403
Combustion **1:** 62, **2:** 300
Complete metamorphosis **2:** 341
Compost pile **1:** 106 [ill.], 107 [ill.]
Composting **1:** 105-121
Compression **4:** 635
Concave **3:** 433
Concentration **3:** 445
Condense/condensation **4:** 729
Conduction **2:** 323
Conductors **4:** 615
Confined aquifer **2:** 308, 310 [ill.]
Coniferous trees **1:** 36 [ill.]

Constellations **4:** 604
Continental drift **4:** 684
Control experiment **1:** 2, **3:** 560
Convection **2:** 325, 326 [ill.]
Convection current **2:** 326, **4:** 684
Convex **3:** 433
Corona **1:** 177
Cotyledon **2:** 265
Crust **3:** 528
Current **4:** 615
Cyanobacteria **1:** 23
Cycles **1:** 175
Cytology **1:** 50
Cytoplasm **1:** 51 [ill.], 51

D

Darwin, Charles **4:** 647, 647 [ill.], 667
Da Vinci, Leonardo **2:** 249 [ill.]
Decanting **3:** 404
Decibel(dB) **4:** 590
Decomposition **1:** 75, 108
Decomposition reaction **1:** 76
Density **1:** 123-138, 124 [ill.], **3:** 542, **4:** 746
Detergent between water and grease **4:** 700 [ill.]
Dependent variable **3:** 560
Desert **1:** 35
Dewpoint **4:** 729
Dicot **1:** 56 [ill.]
Diffraction **2:** 360
Diffraction grating **2:** 363
Diffusion **3:** 445-459
Disinfection **2:** 316
Dissolved oxygen (DO) **1:** 139-158
Distillation **3:** 404
DNA (deoxyribonucleic acid) **1:** 51 [ill.], 52
Domains **3:** 369, 370 [ill.]
Dormancy **1:** 20, **2:** 263
Drought **2:** 232
Drum **4:** 590 [ill.]
Dry cell **2:** 187
Dust Bowl **2:** 232
Dynamic equilibrium **3:** 447

experiment
CENTRAL

E

Ear, inside of human **4:** 592 [ill.]
Earthquakes **1:** 159-174, 160 [ill.], 161 [ill.]
Eclipses **1:** 175-184
Ecologists **2:** 343
Ecosystem **1:** 143, **4:** 748
Electric charge repulsion **3:** 403
Electric eel **2:** 218
Electrical energy **3:** 509
Electricity **2:** 185-201, 302, **4:** 615
Electric motor **2:** 215 [ill.]
Electrode **2:** 186
Electrolyte **2:** 186
Electromagnetic spectrum **2:** 204 [ill.], 205, 357, **3:** 431
Electromagnetic wavelength **2:** 204 [ill.], **3:** 431
Electromagnet **2:** 214 [ill.], 372 [ill.]
Electromagnetism **2:** 203-216, 326, **3:** 371
Electrons **2:** 185, 203, 461, **4:** 615
Electroscope **4:** 619, 620 [ill.]
Elevation **1:** 141
Ellips **2:** 278
Embryo **2:** 263
Endothermic reaction **1:** 61, 63 [ill.], 79
Energy **3:** 509
Enzymes **2:** 217-230
Enzymology **2:** 219
Ephemerals **1:** 37
Epicenter **1:** 160
Equilibrium **4:** 634
Erosion **2:** 231-247
Escher, M.C. illusion **3:** 443 [ill.]
Euphotic zone **3:** 495
Eutrophication **1:** 141
Evaporate/evaporation **3:** 404, **4:** 737 [ill.]
Exothermic reaction **1:** 61, 62, 63 [ill.], 79
Experiment **3:** 560

F

Fault **1:** 159
Fault blocks **1:** 159
Filtration **2:** 316, 404
Fireworks **1:** 78
Flammability **1:** 77
Flight **2:** 249-261
Fluorescence **2:** 360
Focal length **3:** 432, 434 [ill.]
Focal point **3:** 432
Food web **4:** 650
Force **2:** 278
Forest **1:** 35
Fossil fuels **1:** 1, 292
Fourier, Jean-Baptiste-Joseph **2:** 291
Fracture **3:** 533
Franklin, Benjamin **4:** 617 [ill.]
Frequency **2:** 204, **4:** 589
Friction **4:** 615
Frogs **2:** 343 [ill.]
Fronts **4:** 762
Fungus **1:** 23

G

Galaxy **4:** 605
Galilei, Galileo **4:** 603, 604 [ill.]
Genes **1:** 19
Genetic material **4:** 665
Geology **3:** 528
Geotropism **4:** 648
Germ theory of disease **3:** 388
Germination **2:** 263-276
Gibbous moon **1:** 182
Glacier **2:** 293
Global warming **2:** 292
Glucose **3:** 494
Golgi body **1:** 52
Grassland **1:** 35
Gravity **2:** 277-290
Greenhouse **2:** 292 [ill.], 667 [ill.]
Greenhouse effect **2:** 291-306, **4:** 576
Greenhouse gases **2:** 294
Groundwater **2:** 308 [ill.], 309 [ill.]
Groundwater aquifers **2:** 307-321
Growth rings (trees) **1:** 20

Bold type indicates volume number.

general index

H

Halley, Edmond **1**: 176
Heat **1**: 61, 323-339
Heat energy **2**: 323
Helium balloon **3**: 446 [ill.]
Herbivore **4**: 668
Hertz (Hz) **4**: 589
Heterotrophs **1**: 23
High air pressure **4**: 762
H.M.S. Challenger **3**: 542 [ill.]
Hooke, Robert **1**: 49
Hormone **4**: 648, 666
Hot air balloon **2**: 325 [ill.]
Humidity **4**: 745
Humus **1**: 105, 107
Hutton, James **3**: 527, 528 [ill.]
Hydrogen peroxide **2**: 221 [ill.]
Hydrologic cycle **4**: 713, 729
Hydrologists **4**: 730
Hydrology **4**: 729
Hydrometer **3**: 543
Hydrophilic **4**: 699
Hydrophobic **4**: 699
Hydrotropism **4**: 649
Hypertonic solution **3**: 447
Hypotonic solution **3**: 447

I

Igneous rock **3**: 528
Immiscible **1**: 125
Impermeable **2**: 307
Impurities **2**: 316
Incomplete metamorphosis **2**: 341
Independent variable **3**: 560
Indicator **3**: 479
Inertia **2**: 278
Infrared radiation **2**: 291, 326
Ingenhousz, Jan **3**: 493, 494 [ill.]
Inner core **3**: 528
Inorganic **2**: 233
Insulation/insulator **2**: 185, 291,
    4: 615
Interference fringes **2**: 359
Ions **1**: 1, 185, 403, 477
Ionic conduction **2**: 185
Isobars **4**: 762
Isotonic solutions **3**: 447

J

Janssen, Hans **1**: 49

K

Kinetic energy **3**: 509-525
Kuhne, Willy **2**: 217

L

Landfills **1**: 105-121, 108 [ill.]
Langley, Samuel Pierpont **2**: 251
Larva **2**: 341
Lava **3**: 528, 529 [ill.], 683
Leaves **1**: 92, 93 [ill.]
Leeuwenhoek, Anton van **1**: 49, 387
Lens **1**: 49, 50 [ill.]
Lichens **1**: 22 [ill.], 22
Life cycles **2**: 341-356
Lift **2**: 250
Light **2**: 357
Lightening **4**: 618 [ill.]
Light-year **4**: 604
Lilienthal, Otto **2**: 250 [ill.]
Lind, James **3**: 420 [ill.]
Lippershey, Hans **1**: 49
Litmus paper **3**: 479
Local Group, The **4**: 606
Lockyer, Joseph Norman **1**: 178 [ill.]
Low air pressure **4**: 762
Luminescence **1**: 79
Lunar eclipse **1**: 177 [ill.]
    partial lunar eclipse **1**: 178
    total lunar eclipse **1**: 177
Luster **3**: 533

M

Macroorganisms **1**: 106
Magma **3**: 528, 684
Magma chambers **4**: 684
Magma surge **4**: 686
Magnet **3**: 370 [ill.]
Magnetic circuit **3**: 371
Magnetic field **2**: 203, **3**: 369
Magnetic resonance imaging (MRI)
    2: 205 [ill.]
Magnetism **3**: 369-385
Mantle **3**: 528

Manure **1**: 105
Maruia River **4**: 715 [ill.]
Mass **1**: 123, 278
Matter **1**: 123, **4**: 615
Meandering river **4**: 715
Meniscus **4**: 699
Metamorphic rock **3**: 531
Metamorphosis **2**: 341
Meteorologists **4**: 747, 759
Meteorology **4**: 762
Michell, John **1**: 160
Microclimate **2**: 294
Micromes **1**: 106
Microorganisms **1**: 105, 107, 108, **3**: 387-401, 389 [ill.]
Micropyle **2**: 264
Microscopes **1**: 49, 50 [ill.]
Milky Way **4**: 603
Mineral **3**: 527-539
Mixtures **3**: 403-417, 404 [ill.]
Moh's hardness scale **3**: 534
Molecules **1**: 61, **3**: 445
Molting **2**: 341
Monocot **1**: 56 [ill.]
Monument Valley **1**: 37 [ill.]
Moon **1**: 175
Mountain **3**: 530 [ill.]
Mount Tolbachik (volcano) **4**: 685 [ill.]

N
Nanometer **3**: 431
Nansen bottles **3**: 544
Nebula **4**: 604, 605 [ill.]
Neutralization **1**: 3, **3**: 477
Neutrons **3**: 461
Newton, Isaac **2**: 277 [ill.], 277, 357 [ill.], 357, **3**: 511
Niagrara Falls **4**: 716 [ill.]
Nile river **4**: 714 [ill.]
Nonpoint source **2**: 310
Nucleus **1**: 51, **4**: 615
Nutrient **3**: 419
Nutrition **3**: 419-429
Nymphs **2**: 341

O
Oceanography **3**: 541
Oersted, Hans Christian **3**: 370 [ill.]
Optical illusions **3**: 431-444, 440 [ill.], 443 [ill.]
Optics **3**: 431-444
Organelles **1**: 52
Organic **2**: 233
Organic waste **1**: 105
Orion Nebula **4**: 605 [ill.]
Osmosis **3**: 445-459
Outer core **3**: 528
Oxidation **3**: 461
Oxidation-reduction **3**: 461-476
Oxidation state **3**: 461
Oxidizing agent **3**: 463
Ozone layer **2**: 291-306, **4**: 576

P
Passive solar energy system **4**: 576
Pasteurization **3**: 388
Pasteur, Louis **3**: 388 [ill.]
Peaks **2**: 204
Penicillin **3**: 389
Pepsin **2**: 218
Percolate **2**: 307
Permeable **2**: 307
PH **1**: 1, 477-491
Pharmacology **1**: 91
Phases **1**: 182 [ill.]
Phloem **3**: 494
PH levels in U.S. **1**: 5 [ill.]
PH meter **3**: 479 [ill.]
Phosphorescence **2**: 360
Photography **3**: 433 [ill.]
Photosynthesis **1**: 23, 91, 139, **3**: 493-507, 494 [ill.], **4**: 649
Phototropism **4**: 647
Photovoltaic cells **4**: 577
Physical change **1**: 75, 76 [ill.]
Physical properties **1**: 75
Physiologists **3**: 493
Phytoplankton **1**: 139, **3**: 495 [ill.], 495
Pigment **3**: 493

general index

Bold type indicates volume number.

general index

Pitch 4: 590
Plants 1: 92
    Healthy 1: 92 [ill.]
    Unhealthy 1: 92 [ill.]
Plates 1: 159
Platform 4: 635
Point source 2: 310
Poison 1: 1
Polar ice caps 2: 293 [ill.]
Pollination 4: 665, 666[ill.]
Pollution 1: 1
Pores 2: 307
Potential energy 3: 509-525
Potter, Beatrix 1: 22
Precipitation 4: 729
Prisms 2: 358 [ill.], 3: 432 [ill.]
Probes 2: 186
Producer 4: 668
Products 1: 61, 77
Prominences 1: 177
Propellers 2: 256
Properties of light 2: 357-367
Properties of water 4: 697-712
Protein 3: 426
Protists 3: 388
Protons 2: 203, 461, 4: 615
Protozoan 1: 52, 3: 387
Pupa 2: 341

R

Radiation 2: 205, 360
Radicule 2: 265
Radio wave 2: 326
Radiosonde balloons 4: 759
Rain 1: 1-18, 4: 729, 732 [ill.]
Rainforest 1: 35, 38 [ill.], 38
Reactants 1: 61, 77
Reactions 2: 217
Reaumur,Rene Antoine de 2: 217
Recycling 1: 109
Red Sea 3: 543
Reduction 3: 461
Reflection 2: 358, 3: 431
Refraction 3: 431
Relative density 1: 124, 125 [ill.]
Rennin 2: 219
Resistance 2: 188

Respiration 1: 139, 265, 3: 493
Resultant 4: 633, 634 [ill.]
Retina 3: 431
Richter, Charles F. 1: 161, 162 [ill.]
Richter scale 1: 161
Rivers 4: 713-727
Rock 3: 527-539
Roller coaster 3: 518
Roots 2: 264 [ill.]
Runoff 2: 307
Rust 1: 62 [ill.], 462 [ill.], 469 [ill.]

S

Salinity 1: 141, 3: 541-555
Satellites 2: 279 [ill.]
Satellite image 4: 761 [ill.]
Saturated 4: 729
Scars (tree trunk) 1: 20,
Scientific method 3: 557-573
Scurvy 3: 419
Sedimentary rock 3: 529
Sedimentation 2: 316
Seedlings 2: 263 [ill.], 263
Seismic belt 3: 528
Seismic waves 1: 159
Seismograph 1: 161, 4:686
Seismology 1: 161
Semipermeable membrane 3: 445
Sexual reproduction 4: 665
Shapes 4: 633-646
Silt 2: 231
Sirius (star) 4: 605 [ill.]
Smith, Robert Angus 1: 3
Smokestacks 1: 3 [ill.]
Snow 4: 729
Solar cells 4: 577
Solar collector 4: 575, 577 [ill.]
Solar eclipse 1: 175, 176 [ill.]
    Partial solar eclipse 1: 177
    Total solar eclipse 1: 176
Solar energy 1: 4, 4: 575-588
Solar energy plant 4: 578 [ill.]
Solar eruption 4: 576 [ill.]
Solute 3: 403
Solutions 3: 403-417
Solvent 3: 403
Sound 4: 589-602

Specific gravity **1:** 125, **3:** 544
Spectrum **2:** 204 [ill.], 204, 358, 359 [ill.]
Stars **4:** 603-614, 605 [ill.]
Static electricity **4:** 615-632
Streak **3:** 532
Structures **4:** 633-646
Subatomic particles **1:** 123
Substrate **2:** 219
Succulent **1:** 37
Sulfur **1:** 3,
Sun **1:** 175, **4:** 746
Surface tension **4:** 698, 702 [ill.]
Surface water **2:** 309
Suspensions **3:** 403
Symbiosis **1:** 23
Synthesis reaction **1:** 75

T
Taiga **1:** 35
Tectonic plates **4:** 686
Temperature **2:** 323
Thermal energy **3:** 509
Thiamine **3:** 420
Thigmotropism **4:** 649
Tides **2:** 280
Titration **3:** 479
Topsoil **2:** 231
Tornado **4:** 763 [ill.]
Toxic **1:** 1,
Trace elements **3:** 421
Trees **1:** 4 [ill.], 19,
Tropism **4:** 647-663
Troposphere **4:** 745
Tsunami **1:** 173
Tundra **1:** 35
Turbulence **2:** 256
Tyndall effect **3:** 405 [ill.], 406
Tyndall, John **2:** 291

U
Ultrasound **4:** 601 [ill.]
Ultraviolet **2:** 360

Unconfined aquifer **2:** 308, 310 [ill.]
Universal indicator paper **3:** 478 [ill.]
Universal law of gravity **2:** 277

V
Vacuole **1:** 52
Variable **3:** 560
Vegetative propagation **4:** 665-682
Visible spectrum **2:** 357
Vitamin **3:** 420
Volcanos **4:** 683-696, 685 [ill.]
Volta, Alessandro **2:** 186 [ill.]
Voltage **2:** 187
Volta pile **2:** 187
Voltmeter **2:** 186, 189 [ill.]
Volume **1:** 123

W
Waste **1:** 107
Water bug **4:** 699
Water (hydrologic) cycle **4:** 729-743, 730 [ill.], 732 [ill.]
Waterfalls **1:** 140
Water molecule **4:** 698 [ill.]
Water table **2:** 307
Water pollution **1:** 1, 310 [ill.], 311 [ill.], 312 [ill.]
Water vapor **4:** 729, 731 [ill.]
Wave **2:** 204, **4:** 591 [ill.], 592 [ill.]
Wavelength **2:** 204
Weather **4:** 745-758
Weather forecasting **4:** 759-773
Wegener, Alfred **4:** 683, 684 [ill.]
Weight **2:** 280 Wet cell **2:** 187
Wetlands **2:** 310
Wright brothers **2:** 251 [ill.], 251
Wright, Orville **2:** 251
Wright, Wilbur **2:** 251

X
Xanthophyll **1:** 92, **3:** 494
Xerophytes **1:** 37
Xylem **1:** 20, **3:** 494

Bold type indicates volume number.